AF621651

99-ECO-2

Developing Ambient Water Quality Criteria for Mercury: A Probabilistic Site-Specific Approach

by:

Dwayne R.J. Moore, Andrew Pawlisz, and R. Scott Teed
The Cadmus Group, Inc.

G. Mark Richardson
RiskLogic Scientific Services, Inc.

Herbert E. Allen and Sagar Thakali
Department of Civil and Environmental Engineering, University of Delaware

Jennifer C. Gibson, Brian C. Hickey, Jonathan R. Hill, Jonathan Holmes, and Jeffrey J. Ridal
St. Lawrence River Institute of Environmental Sciences

David R.S. Lean
Department of Biology, University of Ottawa

June Crow, David Eskew, George Holdsworth, and Jackie Little
TN & Associates, Inc.

2003

Water Environment Research Foundation

The Water Environment Research Foundation, a not-for-profit organization, funds and manages water quality research for its subscribers through a diverse public-private partnership between municipal utilities, corporations, academia, industry, and the federal government. WERF subscribers include municipal and regional water and wastewater utilities, industrial corporations, environmental engineering firms, and others that share a commitment to cost-effective water quality solutions. WERF is dedicated to advancing science and technology addressing water quality issues as they impact water resources, the atmosphere, the lands, and quality of life.

For more information, contact:
Water Environment Research Foundation
635 Slaters Lane, Suite 300
Alexandria, VA 22314-1177
Tel: (703) 684-2470
Fax: (703) 299-0742
www.werf.org
werf@werf.org

This report was co-published by the following organization. For non-subscriber sales information, contact:

IWA Publishing
Alliance House, 12 Caxton Street
London SW1H 0QS, United Kingdom
Tel: +44 (0) 20 7654 5500
Fax: +44 (0) 20 7654 5555
www.iwapublishing.com
publications@iwap.co.uk

© Copyright 2003 by the Water Environment Research Foundation. All rights reserved. Permission to copy must be obtained from the Water Environment Research Foundation.
Library of Congress Catalog Card Number: 2003104089

WERF ISBN: 1-893664-88-0 IWAP ISBN: 1-84339-637-8

This report was prepared by the organization(s) named below as an account of work sponsored by the Water Environment Research Foundation (WERF). Neither WERF, members of WERF, the organization(s) named below, nor any person acting on their behalf: a) makes any warranty, express or implied, with respect to the use of any information, apparatus, method, or process disclosed in this report or that such use may not infringe on privately owned rights; or b) assumes any liabilities with respect to the use of, or for damages resulting from the use of, any information, apparatus, method, or process disclosed in this report.

The Cadmus Group, Inc.

This document was reviewed by a panel of independent experts selected by WERF. Mention of trade names or commercial products does not constitute WERF endorsement or recommendations for use. Similarly, omission of products or trade names indicates nothing concerning WERF's positions regarding product effectiveness or applicability.

ACKNOWLEDGMENTS

The authors would like to acknowledge Dr. Marti Wolfe, Department of Biological Sciences, California State University, Chico, for the use of her comprehensive mercury bibliographic database, Margaret Stewart, Water Environment Research Foundation (WERF), for her contributions to project management and coordination, and Lori Steeves, Ryan Thompson, and Karen Rowan of The Cadmus Group, Inc. for assistance in editing the document. We would also like to thank the WERF Project Subcommittee (WERF PSC) for its patience and for contributing comments and technical support.

Report Preparation

Principal Investigator
Dwayne R.J. Moore, Ph.D
The Cadmus Group, Inc.

Project Team
Andrew Pawlisz
R. Scott Teed
The Cadmus Group, Inc.

G. Mark Richardson, Ph.D
RiskLogic Scientific Services, Inc.

Herbert E. Allen, Ph.D
Sagar Thakali
Department of Civil and Environmental Engineering, University of Delaware

Jennifer C. Gibson
Brian C. Hickey
Jonathan R. Hill
Jonathan Holmes
Jeffrey J. Ridal, Ph.D.
St. Lawrence River Institute of Environmental Sciences

David R.S. Lean, Ph.D.
Department of Biology, University of Ottawa

June Crow
David Eskew
George Holdsworth
Jackie Little
TN & Associates, Inc.

Project Subcommittee

Robert Berger, Research Council Liaison
East Bay Municipal Utility District

John Knezovich, Ph.D., Chair
NEZ Lawrence Livermore National Labs

Keith Linn
Northeast Ohio Regional Sewer District

Guritno Roesijadi, Ph.D.
University of Maryland

Randall Wentsel, Ph.D.
United States Environmental Protection Agency

Marti Wolfe, Ph.D.
Department of Biological Sciences
California State University, Chico

Water Environment Research Foundation Staff

Senior Research Program Director: Patricia Haddon
Senior Project Manager: Margaret Stewart

ABSTRACT

This project examined the development of ambient water quality criteria (AWQC) for mercury to protect wildlife. Mercury is considered a serious risk to wildlife in many areas. As a result, the Great Lakes Water Quality Initiative and others have developed AWQC. These AWQC have been controversial, however, because 1) the AWQC were single values that did not account for site-specific conditions; 2) derivation of the AWQC relied on a single NOAEL; and 3) the AWQC had an unknown level of conservatism because of reliance on both average and conservative assumptions and uncertainty factors. Rather than develop a single value AWQC for total mercury, this project derives an AWQC model that explicitly incorporates factors controlling bioavailability, methylation rates, and bioaccumulation in the aquatic environment (pH, DOC, sulfate). To derive this AWQC model, field data was collected including numerous water quality parameters and total mercury and methylmercury concentrations in whole body fish tissue from 31 lakes in Ontario and an additional 10 lakes in Nova Scotia. An independent dataset consisting of 51 water bodies in the United States was then used to confirm the validity and robustness of the AWQC model. Next the results of chronic-feeding studies were combined with similar protocols and endpoints in a meta-analysis to derive a dose-response curve for mink exposed to mercury in the diet. Using this approach, an LD_5 or other similar endpoint can be derived that can then be used as the basis for deriving wildlife AWQC. In the final step, a probabilistic risk mode estimated the concentrations of methylmercury in water that would lead to levels in fish sufficient for a 10% probability of exceeding the mink LD_5. This analysis was repeated for various combinations of pH and DOC. The result is an AWQC model for mercury that can be used to protect wildlife in a variety of site-specific conditions.

Keywords: Mercury, Ambient Water Quality Criterion, Wildlife, Site-specific conditions

BENEFITS

- Site-specific ambient water quality criteria (AWQC) for the protection of wildlife.
- Results of an intensive field study examining numerous water quality parameters and fish tissue mercury levels in 31 lakes in Ontario, Canada and 11 lakes in Kejimkujik National Park, Nova Scotia.

TABLE OF CONTENTS

LIST OF TABLES

LIST OF FIGURES

LIST OF ACRONYMS

ANC	Acid Neutralizing Capacity
[Al]	Aluminum Concentration
AWQC	Ambient Water Quality Criteria
APDC	Ammonium Pyrolidine Dithiocarbonate
AH	Aquatic Humus
AA	Atomic Absorption
AAS	Atomic Absorption Spectrometry
AVG	Average
BAF	Bioaccumulation Factor
BrCl	Bromium Chloride
BW	Body Weight
CNS	Central Nervous System
CRM	Certified Reference Material
Cl^-	Chloride Ion
CVAAS	Cold Vapor Atomic Absorption Spectrometry
CVAFS	Cold Vapor Atomic Fluorescent Spectrometry
CONC(s)	Concentration
DDE	Dichlorodiphenyl Dichloroethylene
DIC	Dissolved Inorganic Carbon
DO	Dissolved Oxygen
DOC	Dissolved Organic Carbon
DW	Dry Weight
ECD	Electron Capture Detector
ED_5	5% Effect Concentration
ED_{50}	Median Effect Concentration
FI	Flow Injection
g	Grams
GC	Glass Chromatography
GLC	Glass Liquid Chromatography
GLWQI	Great Lakes Water Quality Criteria
HPLC	High Performance Liquid Chromatography
H	Hour(s)
HCL	Hydrochloric Acid
Hg	Mercury
Hg/L	Mercury Per Liter
HNO_3	Nitric Acid
H_2SO_4	Sulphuric Acid
ICRMS	Inductively Coupled Plasma Mass Spectrometry
K-25	Area In Oak Ridge Reservation, Tennessee
K_{ow}	Octanol-Water Partition Constant
KOH	Potassium Oxide
km	Kilometer

$K_2Cr_2O_7$	Potassium Dichromate
L	Liter
L/kg	Liters per Kilogram
LOAEL	Lowest Observed Adverse Effects Level
M	Molar
MeHg	Methylmercury
MIP-AED	Microwave Induced Plasma Atomic Emission Detection
mg	Milligram
mg/kg	Milligrams per Kilogram
mL	Milliliter
$FeCl_3$	Ferric Chloride
MnO_2	Manganese Oxide
ng/m_3	Nanograms per Cubic Meter
ng/L	Nanograms per Liter
NOAEL	No Observed Adverse Effects Level
NaOH	Sodium Hydroxide
ORR	Oak Ridge Reservation
OMOE	Ontario Ministry of the Environment
pH	Measure of Acidity or Alkalinity
pg/L	Picogram per Litre
QA/QC	Quality Assurance/Quality Control
R^2	Percentage of Variation Explained by Correlation
$SnCl_2$	Stanic Chloride
SPM	Small Particulate Matter
SSIA	Species-Specific Isotope Addition
SRB	Sulfate-Reducing Bacteria
SBMM	Sulfur Bank Mercury Mine
TL	Trophic Level
µg/g	Micrograms per Gram
µg/kg	Micrograms per Kilogram
µL	Microliter
U.S. EPA	United States Environmental Protection Agency
USGS	United States Geological Survey
WK	Week
WW	Wet Weight

Executive Summary

Mercury is considered a serious threat to wildlife in many areas. In recent years a large number of research projects and government policy initiatives aimed at examining and controlling the impact of mercury in the environment have been undertaken (U.S. EPA 1997a,b; U.S. EPA Mercury Action Plan). An important component of the action plan to reduce mercury pollution is the development of ambient water quality criteria (AWQC) to protect aquatic life and wildlife. Existing AWQC for the protection of wildlife are controversial, in particular the scientific rationale behind the wildlife Great Lakes AWQC for mercury (Meyer 1998). Initially, the Great Lakes Water Quality Initiative (GLWQI) proposed a mercury criterion for wildlife of 0.18 ng/L, a level well below the typical mercury analytical detection limit of 200 ng/L and levels that naturally occur in the environment in many locations (Wiener and Spry 1996). The mercury wildlife AWQC was subsequently raised to 1.3 ng/L in the final guidance, without much comment or scientific support. In the Mercury Report to Congress (U.S. EPA 1997a), the authors proposed that the wildlife AWQC be adjusted to 0.64 ng/L total mercury or 0.05 ng/L methylmercury. The updated total mercury value differs from the GLWQI value because the former is national in scope and because of slight differences in how the existing exposure and toxicological information were interpreted (U.S. EPA 1997a).

Three specific factors contributed to the controversy surrounding the Great Lakes Wildlife AWQC for mercury:

1. uncertainty factors and assumptions,
2. single total mercury criteria, and
3. reliance on a no observed adverse effect level (NOAEL).

Each of these factors is briefly discussed below and a recommended approach to address these factors is provided.

Similar to the derivation methodology used to derive AWQC for non-carcinogens for the protection of human health, several uncertainty factors and various assumptions are used in developing wildlife criteria. As a result, the wildlife criterion for mercury has a mix of roughly average assumptions (mink body weight, water and food intake rates), and conservative uncertainty factors (factor of 3 for the extrapolation from a subchronic to a chronic exposure for piscivorous mammals). The net impact of this mix of conservative and average assumptions and uncertainty factors is that the level of protection afforded by the mercury AWQC to wildlife is unknown. Explicitly modeling the uncertainties inherent in the AWQC model for wildlife would be a better approach. The result would be a distribution of possible AWQC values for the protection of piscivorous mammals and birds from which the level of conservatism could be prescribed. Given the uncertainties inherent in deriving an AWQC for wildlife, an appropriate AWQC value, the value that has a 95% certainty of being protective of wildlife, could then be chosen.

The focus on developing a single number for total mercury in the water column is the second major factor. Numerous studies have shown that methylmercury is the form of mercury to which piscivores are exposed through consumption of fish (Grieb *et al.* 1990; Bloom 1992; Stein, Cohen, and Winer 1996) and is far more toxic than inorganic mercury species (Hill and Soares 1984, 1987). Methylmercury concentrations in aquatic systems are a function of total mercury concentrations in water, pH, dissolved organic carbon (DOC), sulfate, redox potential, and other physical and chemical factors (Gilmour and Henry 1991; Gilmour *et al.* 1998; Richardson 1994; Richardson and Currie 1995; Richardson, Egyed, and Currie 1995a,b; Hurley *et al.* 1998). Rather than attempt to develop a single value AWQC for total mercury, it would have been more scientifically defensible to have derived an AWQC equation that explicitly incorporated the factors controlling methylation rates, bioavailability and bioaccumulation in the aquatic environment (pH, DOC, sulfate). This comment also applies to the current Great Lakes and U.S. EPA acute and chronic AWQC for mercury for the protection of aquatic life (U.S. EPA 1985, 1995, 1996b). In its Report to Congress, U.S. EPA (1997a) modified its national AWQC for wildlife to be specific to methylmercury, an improvement over the approach used for the Great Lakes AWQC (U.S. EPA 1995, 1996b). However, the agency still relied on a single Bioaccumulation Factor (BAF) value (the geometric mean for field-measured BAFs in higher trophic levels) and a single conversion factor to convert methylmercury concentrations in water to total mercury concentrations in water. Thus, the revised wildlife AWQC still does not explicitly take account of the physical and chemical factors that control methylation rates and bioaccumulation in aquatic environments.

In this project, an AWQC model was developed where lake-to-lake variations in factors particularly influential to mercury biogeochemistry and fish accumulation are reflected. This model permits the derivation of lake-specific water quality criteria for mercury that are protective of piscivorous wildlife, while acknowledging the heterogeneity of the numerous physical, chemical, and ecological factors among lakes that ultimately influence mercury loading to lakes, methylation, bioavailability and bioaccumulation in fish, and wildlife exposure. Also, the uncertainties inherent in such a model (including measured variability in receptor characteristics of subject wildlife species, reported variability in toxicological response to mercury exposure, variability in lake morphometry and water chemistry) will permit the quantification of the uncertainties present in both the developed model and in the national water quality criterion for mercury derived by the U.S. EPA (1997a).

The reliance on a no observed adverse effect level (NOAEL) from a single, subchronic study that documented neurological damage to mink fed methylmercury in the diet (Wobeser 1973) is the third major factor in the GLWQI controversy. The same approach was taken in the U.S. EPA Report to Congress (U.S. EPA 1997a) in developing its proposed national AWQC for the protection of wildlife. This comment also applies to the approach used for the derivation of the AWQC for the protection of piscivorous birds. In the case of piscivorous mammals, several chronic feeding studies have been published for mink exposed to methylmercury using ecologically more relevant endpoints, such as effects on survival, growth and reproductive fecundity (Aulerich, Ringer, and Iwamoto 1974; Wren *et al.* 1987a,b; Chamberland, Belanger, and Dallaire 1996; Dansereau *et al.* 1999). A better approach would have been to combine the results of these studies, because they followed similar protocols, in a meta-analysis with the purpose of deriving a dose-response curve. Using this approach, one can derive an ED_5 or other similar endpoint that can then be used as the basis for deriving wildlife AWQC. This approach is more scientifically defensible than the one used by U.S. EPA (1995, 1997a) because it relies on all the available evidence and because using an ED_5 is more risk-based than using a lowest observable adverse effect level (LOAEL) (see Suter 1996; Moore *et al.* 1999b). The meta-analysis approach could also be used for the other four at-risk piscivorous wildlife species in the Great Lakes [bald eagles (*Haliaeetus leucocephalus*), herring gulls (*Larus argentatus*), belted kingfishers (*Ceryle alcyon*), river otters (*Lutra canadensis*)]. This approach is used in this project and illustrated using mink as a representative and highly sensitive organism.

The recommendations made above for improving the Great Lakes and U.S. EPA mercury AWQCs for the protection of wildlife are similar to several of the recommendations made by approximately 40 experts who attended a Society of Environmental Toxicology and Chemisty (SETAC) Pellston Workshop entitled "Re-evaluation of the State of the Science for Water Quality Criteria Development," June 25-30, 1998. The purpose of the workshop was to evaluate how ecological risk assessment concepts could be used to develop water quality criteria and methods in a more integrated and efficient process. Specific recommendations that came out of this workshop of relevance to this project included:

- For chemicals where generic water quality characteristics have been shown to have consistent effects on bioavailability, bioaccumulation, and/or toxicity, water quality criteria should be expressed as functions that explicitly incorporate these characteristics.
- There should be less reliance on generic, order-of-magnitude uncertainty factors in deriving AWQC. Alternative approaches include the use of empirically based uncertainty factors (such as incorporating information on the slope of the species sensitivity curve in determining the factor by which the 5th percentile will be divided), probabilistic methods to establish less generic or subjective uncertainty factors from limited empirical data (Richardson and Burmaster 1996), and uncertainty analysis to estimate concentrations that are protective of 95% of aquatic species, or that have a 95% probability of being protective of at-risk wildlife species.
- The results of other lines of evidence (mesocosm studies, field biomonitoring, field validation studies of proposed AWQC) should be used as checks on the appropriateness of proposed AWQC.

Each of these recommendations was incorporated in this study's assessment of existing methods and data development for revising water quality criteria for mercury. The approach to deriving the AWQC for mercury for the protection of wildlife is summarized below in three major components: 1) Field Study, 2) Model Validation, and 3) AWQC Development.

Field Study

A thorough literature search was undertaken to identify sources of data that could be used to develop an AWQC for mercury. A number of papers were identified that contain some of the data required for developing an AWQC for mercury (Gilmour *et al.* 1998; Sorensen *et al.*, 1990); however, these studies fail to report all of the variables necessary. In particular, surveys of fish mercury contamination, with the exception of Sorensen *et al.* (1990), have generally failed to report any measure of mercury (total or methyl) concentration in lake water, because of the rigorous and expensive analytical methods needed to accurately measure total and methylmercury in water and because water mercury concentration is an insignificant predictor (a statistically insignificant regressor) of fish mercury concentration when combined in multiple regression with other variables, particularly DOC concentration. In the case of Sorensen *et al.* (1990), data on total mercury concentration in lake water were not reported for individual lakes, thereby limiting the possible manipulation and extrapolation from that study. A comprehensive 196-page reference guide is available with the online version of this report. For more information on how to access these data, go to our product catalog for this report, Stock no. 99ECO2.

A field research program was undertaken to collect an integrated and consistent dataset suitable for developing a robust ambient water quality criteria model for the protection of wildlife. A complete dataset with many measured variables (Table 1) collected from a large number of lakes is required for this purpose.

Table ES-1. Summary of Variables From the Field Study for Use in Developing the AWQC.

Parameter	Description
Fish Tissue ♦ Mercury ♦ Methylmercury ♦ Length ♦ Weight ♦ Species	♦ Whole body fish tissue samples and few dorsal muscle samples from a minimum of 20 fish specimens (< 20cm length)
Lake Surface Water ♦ Filtered ♦ Unfiltered ♦ Dissolved Organic Carbon ♦ Total Organic Carbon ♦ pH ♦ Conductivity ♦ SO_4 ♦ Turbidity ♦ Color	♦ Sampled at approximately 0.5 m depth beyond the littoral zone of each lake. ♦ Analyzed for total and methylmercury ♦ Filtered and unfiltered samples to determine the contribution of particulate-bound mercury ♦ Sampled only lakes receiving no industrial discharge or municipal effluent discharge

These data were collected from 42 lakes in Ontario (n = 32) and Nova Scotia, Canada (n = 10). Lakes from the Eastern Ontario region that represent a range of water quality characteristics (DOC, pH, alkalinity) known to influence mercury bioavailability were sampled. Lakes with known point sources of anthropogenic mercury loadings (industries, sewage treatment plants, etc.) were excluded. Since the pH of Ontario lakes was comparatively narrow (~7–9), data was also obtained for lower pH lakes from Kejimkujik National Park, Nova Scotia, Canada, where mercury concentrations in loons are among the highest reported (Evers *et al.* 1998) and which have been the sites of recent mercury investigations (summarised by O'Driscoll *et al.* 2001).

Based on the data collected from the field study, mercury in fish was related to total mercury and methylmercury in the water of lakes from which the fish were collected. The authors used these data along with other important variables to develop a multiple regression model that can be employed to predict lake-specific mercury concentrations in fish. Concentrations of methylmercury in water, total mercury in water and total mercury in small fish (4 to 21 cm) are reported. Rather than focussing on a particular species, the authors collected and determined the mercury content of common species in each lake because the goal was to develop a model that could be applied to a wide range of lakes, some of which might contain different prey species.

From the initial dataset (n = 42 lakes) two Nova Scotia lakes were eliminated for which no methylmercury data existed, as well as two other Nova Scotia lakes with too few fish to calculate the Hg concentration of a standard 10 cm fish. The remaining 38 lakes, 31 from Ontario and 7 from Nova Scotia were included in the multiple regression model. The best model to predict mercury concentration in a standard 10 cm fish from lake water chemistry variables was:

$$\mathrm{Log}_{10}(10\ \mathrm{cm}\ C_f) = C - RC_{pH}\ x\ pH - RC_{DOC}\ x\mathrm{Log}_{10}(DOC) + RC_{C_w}\ x\mathrm{Log}_{10}(C_w) \qquad \text{(ES-1)}$$

where C_f (ug/kg) is the concentration of mercury in fish, C (unitless) is a constant, RC_{pH} (unitless) is the regression coefficient for pH, pH (unitless) is a user-entered parameter for the water body of interest, RC_{DOC} (unitless) is the regression coefficient for DOC, DOC (mg C/L) is a user-entered parameter for the water body of interest, RC_{Cw} (unitless) is the regression coefficient for concentration of methylmercury in water, and C_w (pg/L) is the concentration of methylmercury in water. The above regression equation was standardized to 10 cm fish because this is the approximate median size for fish consumed by mink (U.S. EPA 1993).

Stepwise multiple regression analysis identified pH as the most important factor controlling mercury bioaccumulation in fish. The influence of pH alone explained 77.7% of the variation in mercury concentration in fish. Once the variation in pH was accounted for, total mercury concentration in water failed to significantly increase the predictive power of the model. Methylmercury in water and DOC, in contrast, increased the amount of variation explained by the model to 80.4%. This increase, though small, was statistically significant (DOC $p < 0.023$; MeHg $p < 0.03$). Examination of the interaction between DOC, pH, and MeHg indicated that there was no co-linearity between the independent variables ($p > 0.178$ for all possible combinations).

Given the strong influence of lake water chemistry on methylmercury concentrations in both lake water and fish, it is questionable whether a single water quality criterion for mercury in water is adequate to protect fish-eating wildlife. A single criterion applied to lakes with varying water chemistry, food web structure, and watershed characteristics may fail to adequately protect wildlife in some situations while being overly protective in other cases. Given the complexity of the dynamics of the mercury cycle, clearly a more sophisticated approach is warranted.

Model Validation

A validation step was initiated that examined the consistency and potential for generalization of the statistical relationships observed between fish mercury and water chemistry variables for the lakes sampled during the field study. To achieve this validation, an independent dataset was obtained from the U.S. Geological Survey (USGS) with a similar group of variables as collected during the field study. The data were collected from 106 sites from 21 National Water Quality Assessment (NAWQA) basins for the National Mercury Pilot Study. The purpose of the study was to examine relationships between total mercury and methylmercury in water, sediment and fish. Parameters collected in surface water include total and methylmercury, total organic carbon, pH, and sulfate (SO_4). In sediment MeHg, total mercury and % loss on ignition (LOI) was collected. In fish, the weight, length, species, age and mercury content of fish filet data was collected. These data are described in greater detail in Brumbaugh *et al.* (2001) and Krabbenhoft *et al.* (1999). The validation step examines the fish mercury-water chemistry relationships apparent within the independent USGS data and investigates their consistency, or lack thereof, with the data collected from the field study lakes.

The influence of water chemistry on fish Hg accumulation is not expected to be precisely identical from one lake or one group of lakes to another. Variations in physical, geological and biological/ecological factors prevent a precise replication of any particular regression relationship. Also, the necessary differences in the analysis and control of the fish mercury-length relationship between the USGS data and this study's field data preclude a precise replication of the subsequent analysis of water chemistry variables. However, it is important that the nature and direction of influence be the same.

The data obtained from the USGS showed that the influence of water chemistry variables on fish Hg accumulation follows the same general trends as those observed in the field study lakes. As pH declines, fish mercury increases; as DOC declines, fish mercury increases; and as lake water concentrations of methyl Hg increase, accumulation of Hg by fish also increases. Also, lake water concentrations of total Hg were found to have an insignificant influence on average, length-controlled fish Hg accumulation in both groups of lakes. This is consistent with this study's results and Sorensen *et al.* (1990), and is further confirmation that the measurement and use of total Hg concentration in surface water, as an indicator of potential Hg accumulation risk to fish, piscivorous wildlife, and humans, is invalid and inappropriate.

The regression model developed from the field study data (ES-1) predicted fish Hg concentrations in the USGS lakes with reasonable accuracy. Although the models were found to slightly over-predict measured fish Hg concentrations, the error was minor, with the slope of the predicted to measured fish Hg concentrations being 1.117 (versus a desired slope of unity). It is likely that the predictive capability of the models would have been better if there had been more fish from each validation lake (numbers of fish per lake ranged from 1 to 8) and if the individual length and mercury content of each fish had been reported (USGS only reported average length for all fish caught and average Hg for all fish caught (pooled sample)).

These results suggest that the models developed from the field study lakes to estimate fish Hg concentrations may have wide application to predicting fish Hg levels in other lakes, as a function of lake water methyl Hg concentration, DOC concentration, and lake water pH.

AWQC Development

The recent U.S. EPA Mercury Report to Congress (U.S. EPA 1997a) proposed a wildlife AWQC for mercury of 0.64 ng/L total mercury or 0.05 ng/L methylmercury. This wildlife criterion for mercury has a mix of receptor (mink) characteristics that range in their degrees of conservatism and toxicological uncertainty factors of unknown scientific rationale or unquantifiable uncertainty. The net impact of this mix of assumptions and uncertainty factors is that the level of protection afforded by the derived mercury AWQC to wildlife is unknown.

Methylmercury concentrations in fish consumed by piscivorous wildlife are influenced by a myriad of factors including lake water levels of DOC and sulfate, lake pH and alkalinity, food chain length, as well as a number of physical attributes of lakes that influence mercury influx (Brumbaugh *et al.* 2001; Richardson 1994; Richardson and Currie 1995; Richardson, Egyed, and Currie 1995a; Cabana *et al.* 1994).

Although the concentration of methylmercury in lake water is ultimately at the root of methylmercury accumulation in fish and subsequent wildlife exposure through fish consumption, the water concentration of methylmercury, or even total mercury, has been the subject of scant research in relation to fish mercury accumulation. This has resulted, in part, from the relatively low concentrations of total mercury and methylmercury in surface waters, levels generally below routine analytical methods and detection limits until recently. This is coupled with the complex interactions of methylation/demethylation rate (whether in the water column or sediments), with water and sediment chemistry, sediment redox potential, and other factors (Hurley *et al.* 1998; Gilmour *et al.* 1998; Gilmour and Henry 1991; Beijer and Jernelov 1979; Miskimmin, Rudd, and Kelly 1992; Weber 1993; Winfrey and Rudd 1990; Xun, Campbell, and Rudd 1987).

Sorenson *et al.* (1990) have published a study in which water levels of total mercury were collected in addition to the numerous water chemistry, food chain, and lake physical attributes that have been repeatedly shown to influence mercury accumulation in fish. In that study, data on 40 variables collected from 80 lakes in northeastern Minnesota were subjected to biostatistical analysis to ascertain their relative statistical influence on mercury concentrations in northern pike (*Esox lucius*) of 55 cm standardized length. The study found that lake water Hg concentration was an insignificant predictor of fish mercury concentration when combined in multiple regression with DOC and lake pH. Also, simple regression analyses demonstrated that DOC, pH, and zooplankton Hg concentration had a greater influence on fish mercury concentration than did water concentration of total Hg.

The significance of DOC to fish mercury accumulation has been noted by several authors (Wren *et al.* 1991; McMurtry *et al.* 1989). This influence likely relates to the role of DOC in the transport of mercury from watersheds into lakes via precipitation runoff (Lee and Iverfeldt 1991; Mierle and Ingram 1991; Richardson and Currie 1995), its role in complexation and bioavailability of Hg in the water column (Richardson and Currie 1995), and possibly abiotic methylation (Weber 1993). Consistently, DOC is one of the most significant determinants of Hg accumulation in aquatic biota (reviewed by Richardson 1994). The role of pH in fish mercury accumulation may relate to its influence on methylation rate (Gilmour and Henry 1991; Miskimmin, Rudd, and Kelly 1992; Xun, Campbell, and Rudd 1987) and also to its influence on DOC solubility, thereby indirectly affecting Hg bioavailability (Richardson and Currie 1995; reviewed by Richardson 1994). Food chain length has also been shown to be an important determinant of mercury accumulation in top predators of aquatic systems (Cabana *et al.* 1994). The dose-response model derived for mink exposed to mercury is an updated version of the model published previously by Moore *et al.* (1999a). The updated model described below takes advantage of a mink feeding study recently published by Dansereau *et al.* (1999).

For methylmercury, several long-term feeding studies (93 to 183 days) with female mink have been conducted using similar protocols (Aulerich, Ringer, and Iwamoto 1974; Wobeser, Neilsen, and Schiefer 1976ab; Chamberland, Belanger and Dallaire 1996; Dansereau *et al.* 1999). For this project, these studies were combined to produce a toxicity dataset with 16 treatment levels. The endpoint was female mortality that, although not ideal, results in similar effects estimates as occur with reproductive endpoints such as kit biomass or number of surviving kits (Wren *et al.* 1987b). The concentration-response relationship is shown in Figure ES-1.

To convert effect concentrations to doses, the log-logit concentration-response model was combined with the food intake rate for captive female mink. Bleavins and Aulerich (1981) found that captive female mink have a mean food intake rate of 155 g/day (normalized to a 1 kg female mink). The dose-response model is shown in Figure ES-2. The results indicate that the LD_5 for mink exposed to methylmercury in the diet is 80.5 μg/kg bw/day with 95% fiducial limits of 53.6 and 96.8 μg/kg bw/day.

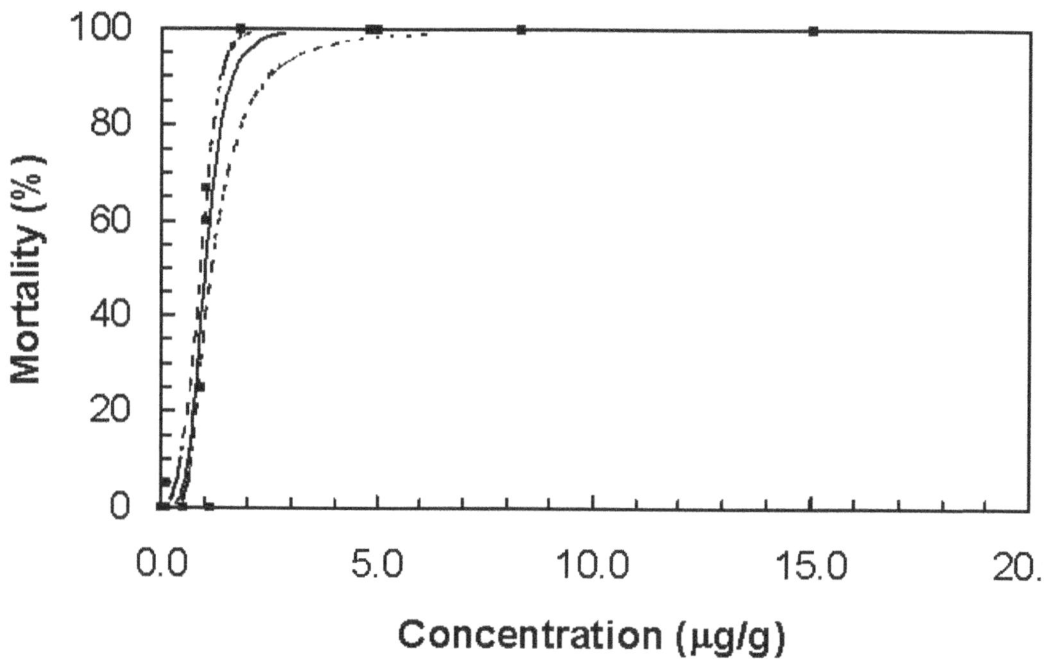

Figure ES-1. Concentration-response Curve with 95% Fudicial Limits for Effects of Dietary Methylmercury on Survival of Female Mink During Chronic Exposures.
Note: Squares indicate original data.

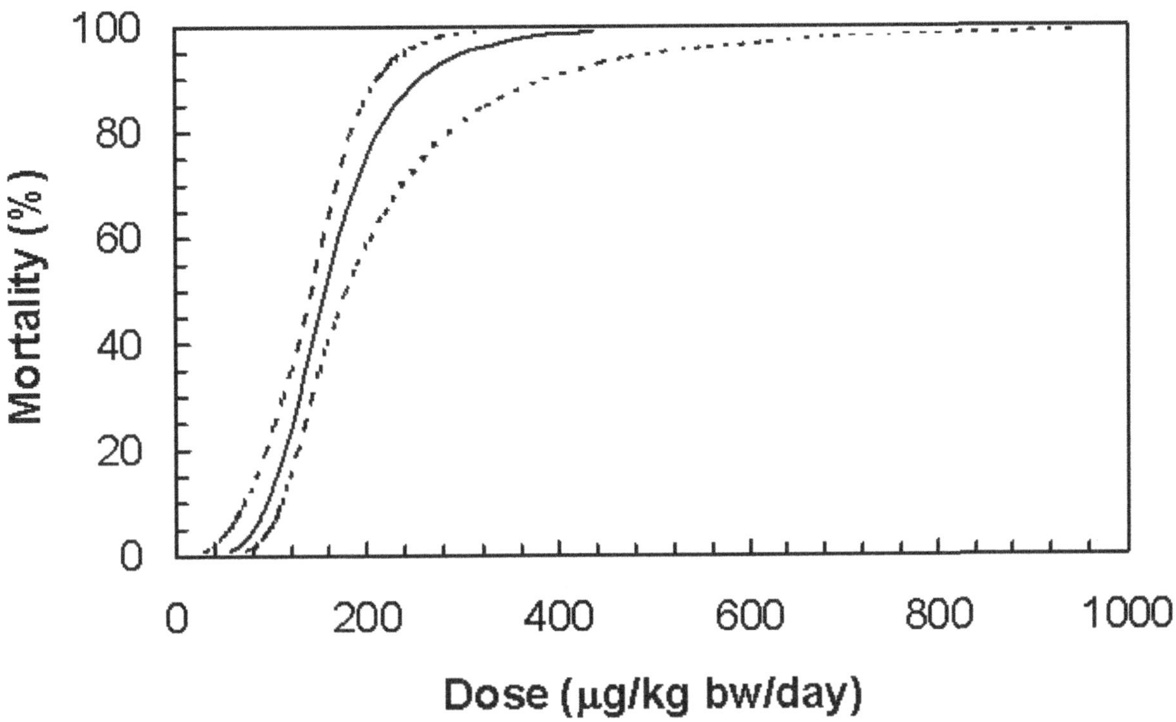

Figure ES-2. Estimated Dose Response Curve with 95% Fiducial Limits for Effects of Dietary Methylmercury on Survival of Female Mink During Chronic Exposures.

Estimating Levels of Mercury in Water Associated with 10% Probability of Exceeding Mink LD_5

A probabilistic risk model was used to estimate ambient water quality criteria for mercury for different combinations of the water quality parameters identified as having a significant influence on mercury levels in fish. Risk was estimated using the following equation:

$$RQ = \left(\frac{MR_{fw} \cdot C_f}{AE_f \cdot GE_f} \right) \div LD_5 \qquad \text{(ES-2)}$$

where RQ is the risk quotient (unitless), MR_{fw} is the metabolic rate of wild female mink (kcal/kg bw/day), C_f is the concentration of mercury in whole fish tissues normalized to 10 cm in length, AE_f is the food assimilation efficiency for mink consuming fish (unitless), GE_f is the gross energy of fish, and LD_5 is the dose causing 5% mortality to female mink (mg/kg bw/day).

The numerator portion of the equation is typical of that used to estimate total daily intake for wildlife (U.S. EPA 1993). The exposure portion of the equation estimates mink exposure to methylmercury, because this form of mercury is by far the most toxic (ATSDR 1989). This requires multiplying the variable for concentration of total mercury in fish (C_f) by the proportion of methylmercury in fish tissues (P_{MeHg}). The above equation also assumes that 100% of the mink diet is fish. This assumption is made for two reasons: 1) in some locations and at certain times of the year mink prey almost exclusively on fish (Alexander 1977) and AWQC should be protective of these situations, and 2) the wildlife AWQC is intended to be protective of other sensitive piscivores, such as river otter, that have a high proportion of fish in their diet (U.S. EPA 1993). To develop an AWQC for mercury, ES-2 must be linked to mercury levels in water, which is done here by substituting the empirical equation from the field study (ES-1) to estimate levels in fish (C_f) from levels of mercury in water.

Figure ES-3 shows the results of a typical analysis, in this case for a pH of 5 and DOC of 12 mg C/L.

In this example, the concentration of methylmercury in water that produced a 10% probability of the quotient exceeding 1 (exposure exceeding the LD_5) was 11 pg/L. The results of the sensitivity analysis for this particular example indicated that seven input variables had rank correlations exceeding ±0.1: regression model constant ($C = 0.56$), metabolic rate for wild female mink ($MR_{fw} = 0.48$), regression coefficient for DOC ($RC_{DOC} = -0.43$), regression coefficient for concentration of methylmercury in water ($RC_{Cw} = 0.26$), regression coefficient for pH ($RC_{pH} = -0.21$), gross energy of fish ($GE_f = -0.20$), and 5% lethal dose ($LD_5 = -0.13$). The results of the sensitivity analysis varied, however, depending on the pH-DOC combination used (Table ES-2). For the regression coefficients for DOC, pH, and MeHg in water, the magnitude of the rank correlations increased as the user-specified values for DOC, pH, and MeHg in water increased, respectively (Table ES-2). The importance of other input variables (gross energy of fish, LD^5, metabolic rate of wild female mink) was less affected by changes in the user-specified parameters.

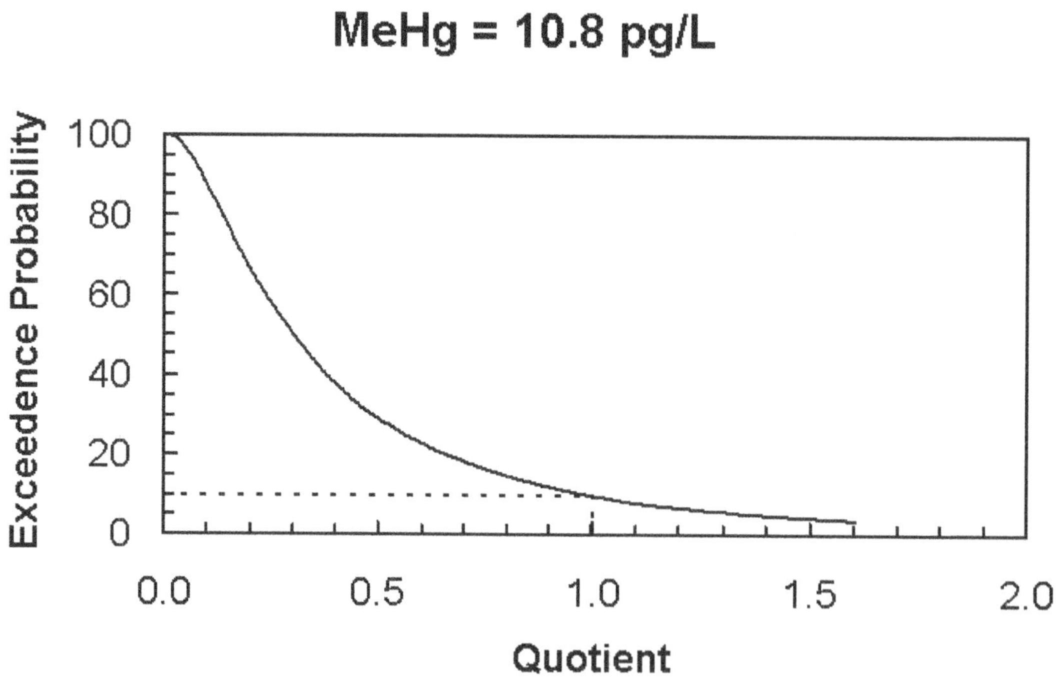

Figure ES-3. Risk Quotient Distribution for pH 5 and DOC Concentration of 12 mg C/L.
The concentration of methylmercury (MeHg) in water that produced a 10% probability of the risk quotient exceeding 1 for wild female mink was 10.8 pg/L.

Table ES-2. Results of Selected Sensitivity Analyses for Different User-Specified Combinations of pH, Dissolved Organic Carbon (DOC) Concentration in Water (mg C/L), and Methylmercury (MeHg) Concentration in Water (pg/L).

Variable	User-Specified Parameters			
	pH=4, DOC=2, MeHg=0.29	pH=9, DOC=2, MeHg=140	pH=4, DOC=14, MeHg=3.6	pH=9, DOC=14, MeHg=1100
C	0.66	0.52	0.59	0.44
MR_{fw}	0.56	0.45	0.5	0.37
RC_{DOC}	–0.14	–0.11	–0.46	–0.35
RC_{Cw}	–0.16	0.49	0.14	0.59
RC_{pH}	–0.2	–0.38	–0.19	–0.31
GE_f	–0.24	–0.21	–0.21	–0.16
LD_5	–0.16	–0.11	–0.13	–0.12

Derivation of the Site-Specific AWQC Equation for Mercury for the Protection of Wildlife

For each pH-DOC combination considered, a concentration of methylmercury in water was determined that corresponded with a 10% probability of exposure exceeding the LD_5 for mink. Using the SAS PROC REG procedure (SAS Institute, Cary, N.C., USA), a multiple regression was performed to derive the AWQC equation for mercury that can be used for any user-specified combination of pH and DOC in water. The dependent variable in this analysis was the concentration of methylmercury in water associated with a 10% probability of exposure exceeding the mink LD_5. This level of risk is used to define the AWQC for mercury that will be protective of wildlife. The list of possible independent variables included: pH, DOC, pH^2, DOC^2, pH^3, DOC^3, $\log_{10}pH$, $\log_{10}DOC$, and pH × DOC. The input dataset included 77 sets of parameters (11 × 7 pH-DOC combinations). In the SAS PROC REG, the MAXR model selection option was used. This option uses forward selection to fit the best one-variable model, the best two-variable model, and so forth. At each level, input variables are switched in and out of the analysis until the combination that maximizes R^2 is determined. The multiple regression analysis to determine the AWQC equation for methylmercury in water indicated that adding input variables to the best four-variable model produced no additional benefit in terms of improving R^2. The R^2 for the four-variable model was 0.9997. Using the best four-variable model, the AWQC equation for mercury to protect wildlife is:

$$Log_{10}AWQC_{MeHg} = -3.748 + 0.348\,pH + 2.491\,(Log_{10}pH) + 1.318\,(Log_{10}DOC) - 0.00204(pH \bullet DOC) \quad \text{(ES-3)}$$

where $AWQC_{MeHg}$ is the ambient water quality criterion for methylmercury in water (pg/L) and DOC is the concentration of dissolved organic carbon in water (mg C/L).

Table ES-3 provides a matrix of pH and DOC values used in the AWQC equation and subsequent AWQC value.

Table ES-3. Derivation of AWQC for Methylmercury for Various Combinations of pH and DOC Using ES-3.

pH	Dissolved Organic Carbon (mg C/L)						
	2	4	6	8	10	12	14
4	0.333	0.8	1.32	1.85	2.39	2.93	3.46
4.5	0.664	1.59	2.6	3.64	4.68	5.7	6.7
5	1.28	3.05	4.97	6.92	8.87	10.8	12.6
5.5	2.41	5.72	9.27	12.9	16.4	19.8	23
6	4.45	10.5	16.9	23.4	29.7	35.7	41.3
6.5	8.07	18.9	30.4	41.8	52.8	63.2	72.8
7	14.4	33.7	53.8	73.6	92.5	110	126
7.5	25.4	59.1	94	128	160	190	217
8	44.4	103	163	220	274	324	368
8.5	76.6	177	278	375	465	546	617
9	131	301	472	633	781	913	1030

Conclusion

The results of the field study and the independent validation analysis indicate the importance of focusing on methylmercury levels in surface waters, not total mercury levels, when managing risks of mercury to wildlife. The site-specific ambient water quality criteria model for mercury to protect wildlife described here is methylmercury specific and provides a highly flexible tool for risk managers. The equation used to generate the criteria can be modified to provide a more conservative level of protection (5% probability of exposure exceeding the LD_5 for mink) or less conservative level of protection (10% probability of exposure exceeding the LD_{10} for mink) as required within specific jurisdictions. Additionally, the AWQC equation allows the risk manager to incorporate as much site-specific information as possible when setting an appropriate AWQC for a particular site. For example, rather than one criterion value being applied that may or may not be appropriate to a specific set of environmental characteristics (high pH, low DOC), the AWQC function described here takes these environmental factors into account on a site-specific basis. Similarly, dose response curves for other piscivorous wildlife species could be generated and applied should other species be more appropriate at a specific site. The end result is the application of appropriate mercury AWQC that will protect piscivorous wildlife on a site-specific basis.

CHAPTER 1.0

INTRODUCTION

Concerns about the human health, social and economic costs of fish advisories for mercury, and the effects of mercury in the ambient environment on wildlife have pushed the mercury issue into the national spotlight in the United States. This attention led to the Mercury Study Report to Congress (U.S. EPA 1997a,b), a document that examines in detail mercury emissions to the environment, fate and levels, risks to human health and the environment, and potential control strategies. Recent and planned U.S. EPA actions will greatly reduce releases of mercury to the environment and possibly mercury exposures. An important component of this action plan is the development of ambient water quality criteria (AWQC) for the protection of aquatic life and wildlife. Such criteria can be used as performance indicators for control strategies being put in place and as the basis for setting effluent limits and discharge reductions at wastewater treatment plants. Because the costs of reducing mercury levels in pre-treatment programs at wastewater treatment plants can be high, municipalities have a strong incentive to ensure that the mercury AWQC developed and revised over the next few years are practical and scientifically defensible.

In 1984, the U.S. EPA derived an acute AWQC (one-hour average) of 2.4 μg/L and a chronic AWQC (4-day average) of 0.012 μg/L (U.S. EPA 1985). The corresponding criteria for saltwater were 2.1 μg/L and 0.025 μg/L, respectively. No criteria were proposed for the protection of wildlife. Much research on the fate and effects of mercury in the aquatic environment took place in the following years. As a result of this new data and the need to derive criteria for the protection of wildlife in the Great Lakes, the U.S. EPA issued a Great Lakes AWQC for mercury of 1.3 ng/L for the protection of Great Lakes wildlife (U.S. EPA 1995, 1996b). The revised acute and chronic criteria for the protection of freshwater aquatic life were 1.7 μg/L and 0.9 μg/L as total recoverable mercury (II), respectively.

The scientific rationale behind the Great Lakes AWQC for mercury has been particularly controversial (Meyer 1998). Initially, the Great Lakes Water Quality Initiative (GLWQI) had proposed a mercury criterion for wildlife of 0.18 ng/L, a level well below the typical mercury analytical detection limit of 200 ng/L and levels that naturally occur in the environment in many locations (Wiener and Spry 1996). The mercury wildlife AWQC was subsequently raised to 1.3 ng/L in the final guidance. Recently, in the Mercury Report to Congress (U.S. EPA 1997a), the authors proposed a wildlife AWQC of 0.64 ng/L total mercury or 0.05 ng/L methylmercury. The updated total mercury value differs from the GLWQI value because the former is national in scope and because of slight differences in how the existing exposure and toxicological information was interpreted (U.S. EPA 1997a).

Several factors contributed to the controversy surrounding the Great Lakes Wildlife AWQC for mercury. First, because the derivation methodology is similar to that used to derive AWQC for noncarcinogens for the protection of human health, several uncertainty factors and various assumptions are used in developing wildlife criteria. As a result, the wildlife criterion for mercury has a mix of roughly average assumptions (mink body weight, water and food intake rates), and conservative uncertainty factors (factor of 3 for the extrapolation from a subchronic to a chronic exposure for piscivorous mammals). The net impact of this mix of conservative and average assumptions and uncertainty factors is that the level of protection afforded by the mercury AWQC to wildlife is unknown. A better approach would be to explicitly model the uncertainties inherent in the AWQC model for wildlife. The result would be a distribution of possible AWQC values for the protection of piscivorous mammals and birds from which the level of conservatism could be prescribed. Given the uncertainties inherent in deriving an AWQC for wildlife, an appropriate AWQC value, the value that would have a 95% certainty of being protective of wildlife, could then be chosen.

The second major factor contributing to the mercury wildlife AWQC controversy was the focus on developing a single number for total mercury in the water column. Numerous studies have shown that methylmercury is the form of mercury to which piscivores are exposed through consumption of fish (Grieb *et al.* 1990; Bloom 1992; Stein, Cohen, and Winer 1996) and is far more toxic than inorganic mercury species (Hill and Soares 1984, 1987). Methylmercury concentrations in aquatic systems are a function of total mercury concentrations in water, pH, dissolved organic carbon (DOC), sulfate, redox potential, and other physical and chemical factors (Gilmour and Henry 1991; Gilmour *et al.* 1998; Richardson 1994; Richardson and Currie 1995; Richardson, Egyed, and Currie 1995a,b; Hurley *et al.* 1998). Rather than attempt to develop a single value AWQC for total mercury, it would have been more scientifically defensible to have derived an AWQC equation that explicitly incorporated the factors controlling methylation rates, bioavailability and bioaccumulation in the aquatic environment (pH, DOC, sulfate). This comment also applies to the current Great Lakes and U.S. EPA acute and chronic AWQC for mercury to protect aquatic life (U.S. EPA 1985, 1995, 1996b). In its Report to Congress, U.S. EPA (1997a) modified its national AWQC for wildlife to be specific to methylmercury, an improvement over the approach used for the Great Lakes AWQC (U.S. EPA 1995, 1996b). However, the agency still relied on a single Bioaccumulation Factor (BAF) value (the geometric mean for field-measured BAFs in higher trophic levels) and a single conversion factor to convert methylmercury concentrations in water to total mercury concentrations in water. Thus, the revised wildlife AWQC still does not explicitly take account of the physical and chemical factors that control methylation rates and bioaccumulation in aquatic environments. In this project, a model that explicitly incorporates information on water quality characteristics influencing the relationship between total mercury levels in water and the levels of methyl-mercury to which piscivorous wildlife are exposed was developed and tested. The model is then used as the basis for deriving mercury AWQC equations, rather than single values, to protect aquatic life and wildlife.

The third major controversial issue for the Great Lakes mercury AWQC was the reliance on a no observed adverse effect level (NOAEL) from a single, subchronic study that documented neurological damage to mink fed methylmercury in the diet (Wobeser 1973). The same approach was taken in the U.S. EPA Report to Congress (U.S. EPA 1997a) in developing its proposed national AWQC for the protection of wildlife. This comment also applies to the approach used for the derivation of the AWQC for the protection of piscivorous birds. In the case of piscivorous mammals, several chronic feeding studies have been published for mink exposed to methylmercury using ecologically more relevant endpoints, such as effects on survival, growth and reproductive fecundity (Aulerich, Ringer, and Iwamoto 1974; Wren *et al.* 1987a,b; Chamberland, Belanger, and Dallaire 1996; Dansereau *et al.* 1999). A better approach would have been to combine the results of these studies, because they followed similar protocols, in a meta-analysis with the purpose of deriving a dose-response curve. Using this approach, one can derive an ED_5 or other similar endpoint that can then be used as the basis for deriving wildlife AWQC. This approach is more scientifically defensible than the one used by U.S. EPA (1995, 1997a) because it relies on all the available evidence and because using an ED5 is more risk-based than using a lowest observable adverse effect level (LOAEL) (see Suter 1996; Moore *et al.* 1999b). The meta-analysis approach could also be used for the other at-risk piscivorous wildlife species in the Great Lakes [bald eagles (*Haliaeetus leucocephalus*), herring gulls (*Larus argentatus*), belted kingfishers (*Ceryle alcyon*), river otters (*Lutra canadensis*)]. The meta-analysis is used in this project and illustrated using mink as a representative and sensitive organism.

The recommendations made above for improving the Great Lakes and U.S. EPA mercury AWQCs to protect aquatic life and wildlife are similar to several of the recommendations made by approximately 40 experts who attended a Society of Environmental Toxicology and Chemisty (SETAC) Pellston Workshop entitled "Re-evaluation of the State of the Science for Water Quality Criteria Development," June 25–30, 1998. The purpose of the workshop was to evaluate how ecological risk assessment concepts could be used to develop water quality criteria and methods in a more integrated and efficient process. Specific recommendations that came from this workshop of relevance to this project included:

- For chemicals where generic water quality characteristics have been shown to have consistent effects on bioavailability, bioaccumulation, and/or toxicity, water quality criteria should be expressed as functions that explicitly incorporate these characteristics.
- There should be less reliance on generic, order-of-magnitude uncertainty factors in deriving AWQC. Alternative approaches include the use of empirically based uncertainty factors (such as incorporating information on the slope of the species sensitivity curve in determining the factor by which the 5th percentile will be divided), probabilistic methods to establish less generic or subjective uncertainty factors from limited empirical data (Richardson and Burmaster 1996), and uncertainty analysis to estimate concentrations that are protective of 95% of aquatic species or that have a 95% probability of being protective of at-risk wildlife species.
- The results of other lines of evidence (mesocosm studies, field biomonitoring, field validation studies of proposed AWQC) should be used as checks on the appropriateness of proposed AWQC.

Each of these recommendations was incorporated in this project's assessment of existing methods and data development for revising water quality criteria for mercury to protect aquatic life and wildlife.

This document begins with a brief chemical characterization of mercury followed by an examination of the analytical methods available for detecting mercury in environmental media (soil, water, and sediment) and tissue. Next, the fate and bioaccumulation of mercury in the environment and the impact of mercury on terrestrial and aquatic biota is examined. Then the authors present their approach to deriving mercury water quality criteria and conclude with a review of remaining data gaps and recommended research for the future.

CHAPTER 2.0

MERCURY SPECIES

Physically, mercury can be in a gas, liquid, or particulate form adsorbed to organic and inorganic particles. Chemically, mercury can exist in three oxidation states: Hg^0 or elemental (metallic) mercury, Hg^+ or mercurous ion (monovalent mercury), and Hg^{2+} or mercury II mercuric ion (divalent mercury) (U.S. EPA 1997a). All three forms of mercury can be found in air, water, and sediments, albeit in different proportions.

In the atmosphere, mercury most often occurs (95% to 99%) as Hg^0 (U.S. EPA 1997a). The remainder exists as Hg^{2+}. Even though Hg^{2+} occurs at a much lower proportion, it is mainly this form that deposits to earth, because Hg^{2+} is much more soluble than Hg^0. Wet deposition can account for as much as 60% to 70% of total deposition of oxidized mercury onto surface waters and land (U.S. EPA 1997a).

In water, mercury occurs mostly as Hg^{2+} in particulate and dissolved form. Usually, methylmercury comprises less than 20% of total mercury in waters near sources and less than 10% of total mercury in waters without a point source (U.S. EPA 1997a). The average proportion of methylmercury in water content is about 8% (U.S. EPA 1997a).

In sediments, Hg^{2+} forms complexes with particles that have a high organic or sulfur content (U.S. EPA 1997a). These complexes are stable and tend to immobilize mercury in the sediment layer. While sediments may act as a sink for mercury, they may also act as a source when biological activity is high enough (U.S. EPA 1997a).

CHAPTER 3.0

Analytical Methods

Mercury occurs in ultra trace amounts in natural environments and is notoriously difficult to analyze for in environmental media due to the high potential for contamination of the samples by outside sources. Contamination during the collection and preparation of the sample and the limitations of analytical methods themselves are two major sources of error in measuring mercury concentrations. Additional confounding factors may include: 1) the interconversion of mercury species during the analytical processes; 2) the use of flawed certified reference materials (CRM); and 3) interference during detection and measurement. Interconversion of mercury species involves the decomposition of methylmercury into elemental mercury and methylation of elemental mercury, which can contribute to an over- or underestimation of measured concentrations depending upon which way the interconversion proceeds during the analysis. Existing CRMs for both methylmercury and total mercury are inadequate for checking the accuracy of the analytical methods under investigation, since they do not represent a wide spectrum of sample matrices, their reported concentrations are based on a limited number of methodologies, and most of the analyses are based on methods that are flawed. For example, though spectrometry methods are generally used to detect mercury in samples, these methods tend to be susceptible to interference and as such may distort the detection and measurement of mercury in samples.

The modern analyses of mercury species (namely inorganic and methylmercury) in water and biological tissues date back to the early 1970s. Most of the methodologies that employ some variation of the Westoo Method (Westoo 1967) for sample treatment and the Hatch and Ott Method (Hatch and Ott 1968) for detection, are reliable.

This section provides a brief summary of the analytical methods and practices used in the detec-tion and measurement of mercury in environmental samples. A full review and description of the existing analytical and detection methods is presented in Appendix A.

3.1 Fish Tissue

Total mercury concentrations in fish tissue are generally in the part per billion (ug/kg) range. Consequently, the analytical detection limit of most standard techniques is adequate to characterize mercury concentrations. However, it is essential that the tissues be handled with clean-room gloves, and stainless steel dissecting implements under trace-metal clean conditions and frozen into acid-cleaned Teflon® vials prior to analyses (Bloom *et al.* 1994; Bloom 1995).

Two generally reliable procedures include:

1. For total mercury in fish tissue, hot acid digestion (7:3 HNO_3/H_2SO_4), followed by $SnCl_2$ reduction, dual amalgamation, and cold vapor atomic fluorescent spectrometry (CVAFS) (Bloom 1992);
2. For methylmercury in fish tissue, alkaline (1:3 KOH/methanol) digestion followed by aqueous phase ethylation, then cryogenic gas chromatography (GC) separation, and CVAFS detection (Bloom 1989). Other derivativizations, particularly phenyl, can be substituted for the ethylation.

3.2 Water

Using clean techniques, the concentration of mercury in water bodies has been found to be in the part per billion to sub part per trillion range in natural waters and in the 1–200 ng/L in effluent waters (Hunt 1998). At such levels, any contamination or loss of the analyte will have a significant effect on the quality of the data. Generating reliable mercury water quality data requires clean conditions during the entire analysis process. The samples should be collected by immersing acid-cleaned Teflon® bottles using clean-room gloves that are changed between samples (Bloom *et al.* 1994). For sample collection, polyvinyl chloride Niskin type sampling bottles (Bothner and Robertson 1975) and Teflon®-coated Go-Flo sampling bottles (Bloom and Crecelius 1983) have also been used. Samples should be frozen immediately. Then they should be acidified to 1% with ultrapure acid [low mercury HCl (< 5 pg Hg/mL)] prior to analyses (Bloom *et al.* 1994).

All lab equipment must be cleaned using hot-oxidation or similar technique (Bloom and Crecelius 1983). The bottles are filled with a mixture of the sample water, 2% HNO_3, and 2% hot oxidizing reagent. Glass stoppers and small pieces of equipment are placed into a large beaker of oxidizing solution. They are then heated to 80–90°C in a hot water bath and kept at this temperature for an hour. After the appearance of MnO_2 (3–5 hours depending on the temperature), the contents are discarded, and all surfaces rinsed with 12% NH_2OH•HCl, followed by deionized water, to remove MnO_2. The bottles are then placed into 6N reagent grade HNO3 for one week. After rinsing in de-ionized water, the hot oxidation procedure is repeated. The sample containers are finally rinsed with the sample water and filled as described above. A reliable analysis of total mercury in water includes: Leaching and oxidizing all forms to Hg(II) with acidified BrCl, reducing with NH_2OH•HCl and $SnCl_2$, purging with N_2, collect on a gold (Au) trap, and measuring by cold vapor atomic absorption spectrometry (CVAAS) (Robertson *et al.* 1987). A laboratory specifically designed for analyzing mercury with clean facilities like the ones proposed by Boutron (1990) or Ferrari, Moreau, and Boutron (2000) should be used. The clean laboratory should be positive pressure with non-laminar flow, have a Hg laminar flow clean bench, and clean garments and gloves should be used. The clean benches should be made of polyethylene and washed with acidified pure water. Use of high efficiency particulate (HEPA) filters for particulate Hg elimination and charcoal filters to eliminate gaseous Hg is critical. Flushing a 700 m^3/h air flow through an active charcoal filter, a prefilter (92% efficiency) and finally on a HEPA filter (99.999% efficiency) has been shown to reduce the gaseous Hg in the clean bench area to 2.2 ± 0.6 pg/L. Ultrapure reagents are prepared with water using a series of activated charcoal and mixed bed ion exchange resin columns (Ferrari, Moreau, and

Boutron 2000). Reliable detection systems include element-specific CVAAS, CVAFS, or the more modern species-specific ICP-MS. ECD data suffers from significant interference, and even the ordinary flameless AAS have reportedly failed to detect mercury species associated with particulate forms (Sanemasa *et al.* 1976).

3.3 Analytical Method Summary

Relatively high tissue mercury concentrations in biological tissues make most of the standard analytical protocols currently in use precise and accurate. The same cannot be said about the ambient and effluent water samples which have a significantly greater risk of contamination during sampling and handling (Colman and Breault 2000). Thus, water quality data produced without clean conditions should be used with caution. In particular, mercury analyses of water before the mid- to late-1980s should be treated with suspicion. The importance of clean techniques can be seen in the analysis of mercury in Vandercook Lake using progressively cleaner sample collection and handling techniques (Fitzgerald and Watras 1989), where the following Hg concentrations in pM were reported: 1983 (~1200); 1985–1986 (~300) and 1986 (2.5 ± 0.2). Total Hg appears to be in the range 1 to 10 ng/L in most areas. Humic lake waters may have levels up to 15 ng/L.

All the toxicity studies described in the remainder of the document have used the ultra-clean technique in the analysis of mercury during exposure and during tissue analysis. This QA/QC step was taken to facilitate comparison between studies and help ensure that the total mercury or methylmercury concentrations reported are reasonably accurate and precise.

CHAPTER 4.0

FATE AND BIOACCUMULATION OF MERCURY IN THE ENVIRONMENT

Mercury (Hg) from both natural and anthropogenic sources is ubiquitous in the environment. In its elemental form, Hg^0, it circulates in the earth's atmosphere, becomes widely dispersed, and slowly oxidizes to inorganic Hg (Figure 4-1). The process of methylation, mostly by anaerobic bacteria in marine and freshwater sediments, results in the conversion of inorganic Hg to organic forms, of which methylmercury (MeHg) is the most common.

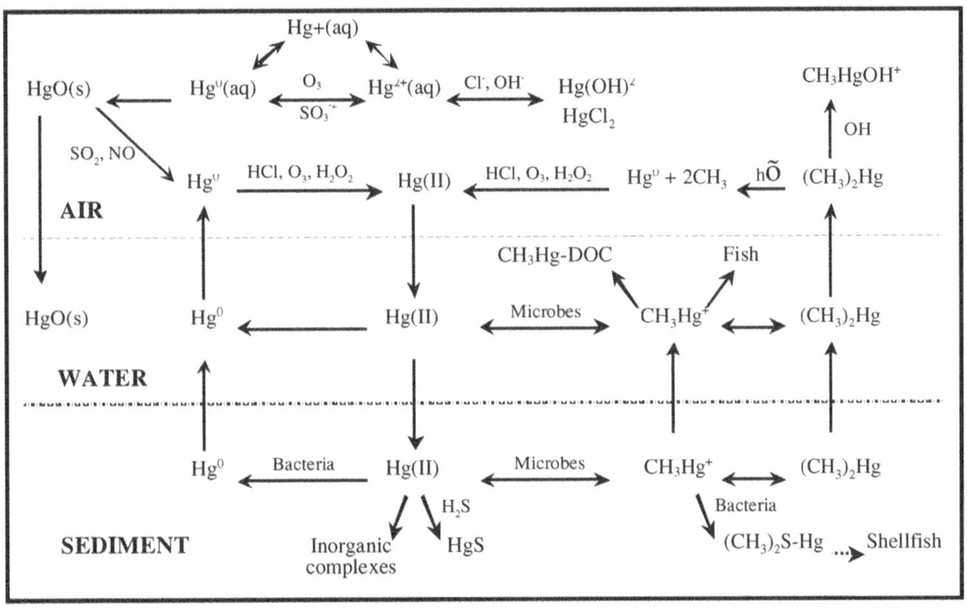

Figure 4-1. Potential Transformations of Hg in Air, Water, and Sediment.

Methylmercury accounts for most of the Hg found in fish that can ultimately serve as food sources for humans. By contrast, inorganic forms of Hg are not as readily absorbed from the gastrointestinal tract and thus are not as highly concentrated in animals as organic forms. Because of: 1) the larger area of the earth's surface that is covered by water, 2) the degree of concentration of MeHg up the food chain (bioaccumulation), and 3) the extent to which some individuals rely upon fish protein as a dietary staple, the potential for exposure to MeHg is widespread among several human subpopulations and wildlife (U.S. EPA 1997b). A dramatic demonstration of the issues involved in bioaccumulation of Hg from aquatic environments is provided by the well-documented poisoning incident at Minamata Bay, Japan, in 1974. In this case, coastal dwellers became sick and died as a result of eating shellfish containing Hg that had been discharged from a local industrial facility (D'Itri 1972; Tsubaki and Takahashi 1986). The shellfish had accumulated the element from lower trophic levels as MeHg, which, with its high solubility and ready capacity to pass across biological membranes, had bioaccumulated during its passage through the food chain and ultimately to man. The Minamata Bay incident is a striking example of how bioaccumulation of MeHg at lower trophic levels can have toxicological conse-quences to organisms linked in a series of predator/prey relationships. Analogous consequences of bioaccumulation leading to poisoning might be expected to occur in other trophic forms that represent the apex of an aquatic food chain in which bioaccumulation of MeHg takes place.

Both the U.S. EPA Mercury Report to Congress (1997a) and the U.S. EPA Great Lakes Water Quality Initiative (1993a,b) have collated and researched the scientific literature relating to the bioavailability and bioconcentration of Hg, based, for the most part, on information that was current to about 1995 or 1996. This discussion, with its emphasis on reports of studies published in the primary literature during the last three years, updates these comprehensive texts. The focus is on drawing together a subset of the more recent information documenting the quantitative impact of biotic and geochemical factors on the passage of MeHg through the food chain. In addition, the emphasis on reports of recent studies ensures that the experimental findings and scientific insights to emerge have been derived through the use of acceptable experimental protocols and analytical techniques. Typically, this involves the use of ultraclean extraction and analytical techniques.

Water quality and related geochemical parameters that affect bioavailability, bioaccumulation, and biomagnification of MeHg by phytoplankton, zooplankton, benthic invertebrates, and small fish are discussed below. Summary tables of bioaccumulation studies examining lower and higher trophic level organisms are found in Appendix B. Much more information exists on the fate and bioaccumulation of mercury in fish, relative to other taxa, which may reflect the research conducted as a result of the Minimata Bay and similar incidents that affected human health.

4.1 Factors Affecting Mercury Bioconcentration and Biomagnification

4.1.1 Mercury Methylation

Biotic routes for the generation of methylmercury are provided by sulfate-reducing bacteria (SRB) and obligate anaerobes present in surficial sediments and the anoxic waters that overlay them. The methylation of inorganic mercury by SRB may be the dominant route for the production of organic forms of mercury such as MeHg. In general, the methylation rate is dependent on many water quality parameters, such as temperature, dissolved oxygen (DO) content, DOC concentration, inorganic Hg concentration, pH, and the concentration of nutrients necessary for microbial growth. The uptake and bioaccumulation of MeHg by the various trophic levels represented in the pelagic food web will ultimately depend on the rate of microbial conversion of inorganic Hg to MeHg.

The sediment surficial layer (the first few centimeters of anoxic sediment) is a primary site for the production of MeHg by SRB. Sediments with a high content of decomposing organic matter are locations where enhanced rates of methylation may be expected, for example, soils and peatlands that have been flooded to provide reservoirs. Reservoirs have been extensively investigated to identify the factors that contribute to increased concentrations of MeHg in fish. Hall, Rosenberg, and Wiens (1998) compared MeHg concentrations in aquatic insects before and after the flooding of a small wetland lake. They observed a three-fold increase in MeHg in predatory insects. Earlier studies at the same lake, cited by the authors, reported a 20-fold increase in water MeHg, and a two- to three-fold increase of MeHg in fish after flooding. According to Boudou and Ribeyre (1997a,b), sediments provide a major storage compartment for Hg, containing up to 99% of the metal in a typical forest lake. MeHg produced by microbial activity in sediment or anoxic water may diffuse into the water column, or, in the case of a stratified lake, be released from the anoxic waters of the hypolimnion during fall mixing. At this point, MeHg is available for uptake by biota.

Although methylation is thought to occur mainly in sediment and adjacent anoxic waters, there is circumstantial evidence cited by Morel, Kraepiel, and Amyot (1998) indicating that MeHg may be produced in the oxic waters of very deep lakes, such as the Great Lakes. Transport of MeHg from the depths of these lakes to the surface is thought to be unlikely, yet there are substantial concentrations of MeHg present in the surface waters of the Great Lakes that are likely to be of bacterial rather than atmospheric origin. The mechanisms are not well understood for either microbial or abiotic methylation of inorganic Hg in either oxic waters or sediments.

As reported by Cleckner (1999), periphyton communities may also provide a site for MeHg production. In Cleckner's experiment in the Florida Everglades (1999), filtered surface water and periphyton samples were collected using clean techniques. To determine the rate of methylation, aliquots of periphyton were spiked with $^{203}HgNO_3$ and then incubated *in vitro* in water collected from the site. The MeHg that evolved during incubation was trapped and extracted, and the activity of ^{203}Hg counted. High rates of methylation were observed for the periphyton from eutrophic areas. Periphyton from less eutrophic areas was shown to be a less active methylator. Periphyton from eutrophic areas was shown to support photosynthetic, sulfur-cycling bacteria that may, in addition, have the capacity to directly methylate inorganic Hg. If the observations made by Cleckner (1999) are confirmed, these findings are important because the periphyton and its associated diatomaceous communities provide the base of the food web in the Everglades and possibly other wetlands.

In tropical lakes, the roots of floating macrophytes, such as *Salvinia sp.*, provide a site for methylation (Guimaraes *et al.* 1998). The rate of methylation associated with the roots of floating vegetation was determined to be a factor of nine higher than in the underlying sediments. Thus, mats of floating macrophytes appear to provide a readily available source of MeHg. In the same study, the authors assessed the distribution of $^{203}HgCl$ spiked into the water column above an undisturbed core of open-lake sediment. The sediment was incubated for three days, after which ^{203}Hg was detected at sediment depths of 14–16 cm, depths at which chironomid larvae are found. MeHg was present only in the surficial layer (0–2 cm) of sediment. Bioturbation may play an important part in the cycling of MeHg and inorganic Hg between sediment and the water column. The authors suggested that the formation of MeHg may be dependent on the progressive penetration of ^{203}Hg into the core sediment.

4.1.2 Uptake Mechanisms

In a review of the chemistry of Hg in oxic waters, anoxic waters, and sediments, Morel, Kraepiel, and Amyot (1998) describe the first key event in the methylation of inorganic Hg as the passage of inorganic Hg into the bacterial cell. The biotransformation of inorganic Hg to MeHg and the subsequent transfer of MeHg to the base of the food chain (the first trophic level) are therefore dependent on initial microbial uptake of inorganic Hg. An array of complex interactions between ionic and non-ionic forms of Hg complicate understanding of the kinetics of Hg uptake into the cell, although several lipophilic components and complexes appear to have the capacity to diffuse passively across the plasma membrane. Thus, inorganic Hg, in the form of a covalently bound complex, can passively diffuse across the cell membrane into the cytosol of the microorganism (Morel, Kraepiel, and Amyot 1998). Several covalently bound forms of inorganic Hg have low octanol/water partition coefficients (K_{ow}), yet are lipophilic. In anoxic, low pH sediments and waters, $Hg(HS)_2$ or perhaps polysulfide complexes (of the general formula HgS_n) may be the covalently bound forms of Hg that partition into bacteria (Morel, Kraepiel, and Amyot 1998). In anoxic waters, inorganic Hg may enter a bacterial cell or diatom as $HgCl_2$. $HgCl_2$ has a K_{ow} value of 3.3, is lipid soluble and is capable of rapid diffusion through the lipid bilayer of a cell membrane. The dominance of any one type of inorganic Hg chloro-complex in water is a function of the prevailing chloride concentration and pH. As pH and chloride concentration increase toward the values observed for seawater, charged inorganic mercury species become dominant. It is unlikely that the charged species of inorganic Hg chloro-complexes will passively diffuse across the cell membrane.

Barkay, Gillman, and Turner (1997) investigated the effects of DOC and salinity on the bioavailability of Hg to bacteria, using a mer-lux bioindicator. The characteristic of Hg resistance was coupled to luminescence (light-emitting capability) in a strain of *Escherichia coli* to create the bioindicator. After modification, the *E. coli* was enabled to emit light in proportion to bioavailable inorganic Hg. As DOC concentrations increased, a decrease in inorganic Hg-dependent light emission was observed, an effect particularly marked at pH 7. The authors suggested that DOC is a controlling factor for inorganic Hg bioavailability. Concentrations greater than 1 mM Cl^- also decreased the bioavailability of inorganic Hg to the test organisms, causing a decrease in luminescence. As chloride concentration increased the charged chloro-mercury species became predominant, decreasing bioavailability. In contrast to the results of a previous study cited by the authors, $Hg(OH)_2$ and HgClOH were found to be bioavailable.

MeHg may enter the cell of a microorganism as an uncharged chloride species CH_3HgCl, with passive uptake possibly being the major route for bioaccumulation of both inorganic Hg and MeHg in phytoplankton. However, recent studies by Watras *et al.* (1998) provide data that are inconsistent with the hypothesis that passive transport of MeHgCl plays a controlling role in trophic transfer. The authors describe the positive correlation between calculated aqueous concentrations of $MeHg^+$ or MeHgOH and seston MeHg. This suggests an active transport of these species into bacteria and phytoplankton, as MeHgOH is known to diffuse more slowly across the cell membrane than does MeHgCl. No correlation was observed between calculated values for MeHgCl in water and seston.

Also, as reported by Morel, Kraepiel, and Amyot (1998), some Hg-resistant bacteria may actively transport inorganic Hg across the microbial cell membrane. In the presence of elevated concentrations of inorganic Hg, bacteria that carry the mer-operon have the ability to synthesize

MerT (membrane transport) protein. As its name suggests, the MerT protein affects the active transport of inorganic Hg into the cell. At lower inorganic Hg concentrations, the passive route into the cell may be dominant.

4.1.3 Biological Factors

There are numerous biological factors that can affect the bioconcentration and bioaccumulation of mercury into biota. As previously discussed, the presence of methylating bacteria is particularly important to generating methylmercury, which is subsequently available to biota. Other biological factors are discussed below.

A considerable bibliography exists on how and to what extent the size and age of fish predators and prey impact the bioaccumulation of MeHg. For example, from older studies, the single best predictor of the concentration of Hg in fish is fish length, due to its strong association with exposure duration (fish age) (Huckabee, Elwood, and Hildebrand 1979). Therefore, interlake comparisons of fish Hg contamination should standardize fish Hg concentration for fish length (McMurtry *et al.* 1989; Wren *et al.* 1991; Sorensen *et al.* 1990). Among more recent studies, Munn and Short (1997) found that Hg concentrations were positively correlated with age, weight, and length of walleye (*Stizostedion vitreum*) collected from three reaches of Lake Roosevelt, a reservoir on the Columbia River. Walleye were separated into four length categories, 25–33 cm, >33–41 cm, >41–48 cm, and >48–56 cm. Concentrations of Hg in composite filets from eight fish in each size category ranged from 0.11 to 0.44 mg/kg and were significantly correlated with mean total length, mean weight, and mean age. A curvilinear regression best described the relationship between age and Hg concentration, with Hg concentration leveling off in fish over 3 years old. Condition factor (K), defined as $K = 10^3 \times$ (ww in g)/(total length in cm)^3 was inversely correlated with Hg concentration. Average walleye Hg concentrations did not vary among the three sampling reaches and did not reflect differences in sediment Hg concentrations reported separately by Bortleson *et al.* (1994). However, no conclusions regarding bioconcentration factors can be drawn from this work, because water Hg and MeHg concentrations and prey Hg concentration were not reported.

Neumann and Ward (1999) investigated total Hg concentrations in fish species from different trophic levels in two impoundment lakes in Connecticut. Largemouth bass (*Micropterus salmoides*), chain pickerel (*Esox niger*), black crappie (*Pomoxis nigromaculatus*), and bluegill (*Lepomis macrochirus*) were collected from Pickerel Lake (35.9 ha, mean depth 1.8 m), and largemouth bass, smallmouth bass (*M. dolomieui*), yellow perch (*Perca flavescens*), and bluegill were collected from Lake Lillinonah (769 ha, mean depth 7.6 m). Statistically significant positive linear regressions were found between total Hg concentration in muscle and both age and length for each species in both lakes. The total Hg concentrations were consistently higher in the top-level predator species (largemouth bass, smallmouth bass, and chain pickerel) than in the intermediate-trophic-level fish species (black crappie and yellow perch), and the lowest concentrations were observed in the low-trophic-level representative (bluegill in this study). The relative differences in total Hg concentrations between bluegill and the top-level predator species increased with age (Table 4-1). In general, the differences between the top-level predator species and the intermediate-trophic-level fish tended to decrease as the fish matured. This may represent a shift in diet from invertebrates to fish as the intermediate level fish matured.

Table 4-1. Effect of Age on Total Mercury Concentrations (µg/g ww (SD)) in Muscle Tissues of Fish Species from Two Connecticut Lakes.

Species	Trophic level	Age (years)			
		2	3	4	5
		Pickerel Lake			
Largemouth bass	Top	0.26	0.31	0.38	0.47
Chain pickerel	Top	0.24 (1.1)	0.30 (1.0)	0.36 (1.1)	0.44 (1.1)
Black crappie	Intermediate	0.07 (3.7)	0.10 (3.1)	0.14 (2.7)	0.21 (2.2)
Bluegill	Low	0.11 (2.4)	0.12 (2.6)	0.13 (2.9)	0.14 (3.4)
		Lillinonah Lake			
Largemouth bass	Top	0.33	0.46	0.64	0.89
Smallmouth bass	Top	0.39 (0.8)	0.51 (0.9)	0.68 (0.9)	0.91 (1.0)
Yellow perch	Intermediate	0.23 (1.4)	0.34 (1.4)	0.50 (1.3)	0.74 (1.2)
Bluegill	Low	0.17 (1.9)	0.20 (2.3)	0.24 (2.7)	0.28 (3.2)

Source: Neuman and Ward 1999.

Henry *et al.* (1998) measured the total Hg concentration in smallmouth bass in a remote lake on the Upper Peninsula of Michigan that experienced minimum impact from anthropogenic sources. Fumee Lake has a maximum depth of 4 m and is fed by several surface springs and artesian wells. The average specific conductance is 272 µmhos/cm, average pH is 8.41, average hardness is 152 mg $CaCO_3$/L, and the average alkalinity is 128 mg $CaCO_3$/L. Fish were separated into size categories of <20 cm, 20–30 cm, and >30 cm, and whole fish were homogenized for analysis. In contrast to the findings of most other studies, the Hg concentration decreased as fish length and weight increased (Table 4-2). A major difference in the values reported by Henry *et al.* (1998) was that the whole body was homogenized for analysis, whereas most studies have used muscle fillets.

Table 4-2. Total Mercury Concentrations in Different Size Classes of Smallmouth Bass from Remote Fumee Lake with Minimal Anthropogenic Inputs.

Length (cm)	Average Weight (g)	Mercury Concentration (µg/kg ww)
0–20	67	303
20–30	268	202
>30	1025	162

Source: Henry *et al.* 1998.

Park and Curtis (1997) compared Hg concentrations in muscle tissue of largemouth bass, bluegill, and crappie taken from two reservoirs in Oregon. The former Black Butte Hg mine was present in the watershed of Cottage Grove Reservoir, whereas the Dorena Reservoir has no history of mining activity. The average basin sediment concentration (mean ± SD) in Cottage Grove Reservoir (0.67 ± 0.41 µg/g dw) was 3.7-fold higher than in Dorena Reservoir (0.18 ± 0.41 µg/g dw). Concentrations of Hg were higher in largemouth bass from Cottage Grove

Reservoir (0.31–0.96 µg/g ww) and were approximately double the concentration range found in largemouth bass from Dorena Reservoir (0.11–0.52 µg/g ww). A positive, linear correlation was observed between largemouth bass age and muscle Hg concentrations when fish taken on the same sampling date were compared. In addition, Hg concentrations in the same age group of fish varied by up to 100% between samples collected in June in comparison to September. However, the cause of such large seasonal fluctuations remains undefined. A positive linear correlation between age and Hg concentrations was also observed for black crappie in Cottage Grove Reservoir. No statistically significant correlations between largemouth bass age and Hg concentrations were observed in the less polluted Dorena Reservoir. Also, in contrast to most other studies, the concentrations of Hg in bluegills in Cottage Grove Reservoir (0.45–1.1 µg/g ww) were higher than those observed in largemouth bass for fish from two years old up to five years old. Unfortunately, concentrations of Hg in the water were not reported, so BAFs cannot be calculated. However, Hg transfer between low trophic level fish and top-level predators appears to have resulted in little additional bioaccumulation of Hg.

Rose *et al.* (1999) surveyed Hg concentrations in sediments, water, and muscle of fish from a wide range of trophic levels in 24 "least-impacted" lakes in Massachusetts. Largemouth bass are piscivorous predators, but they include invertebrates and amphibians in their diets. Yellow perch are omnivores with insects, invertebrates, and fish in their diet. Brown bullheads (*Ameirus nebulosus*) are bottom-feeding omnivores. The selected lakes were located in three ecological subregions characterized by differing geology, forest types, and water characteristics. Average muscle Hg concentrations were lowest in brown bullhead (0.15 mg/kg), intermediate in yellow perch (0.31 mg/kg), and highest in largemouth bass (0.39 mg/kg). The size range of fish selected for analysis in each lake was limited to 20–25 cm for brown bullhead and yellow perch and to 30–36 cm for largemouth bass. The larger size range for largemouth bass was selected because 30.5 cm is the legal kreel size limit. Thus, the potential for evaluating relationships between fish size and Hg bioaccumulation was limited and a statistically significant positive correlation between muscle Hg concentration and size was observed only in largemouth bass. Statistically significant relationships were not observed between sediment Hg or selenium concentrations and muscle Hg concentrations in any species. Low water pH was correlated with elevated Hg concentrations in both brown bullhead and yellow perch, but not in largemouth bass. Rose *et al.* (1999) suggested that species where Hg concentrations vary greatly with size (such as large-mouth bass) might not be good choices for evaluating the effects of environmental variables. The small differences between the average Hg concentrations in the species examined in this study suggest that bioaccumulation between fish trophic levels is limited for the relatively uncontaminated lakes in this study.

Svobodová *et al.* (1999) surveyed Hg concentrations in nine fish species in two reservoirs in the Czech Republic. The Orlík Reservoir receives water from the entire watershed of the Otava and Vltava Rivers, while the Kamyk is a smaller reservoir that serves as an equilibration reservoir directly downstream of the Orlík Reservoir. Fish from six sites along the two rivers and two reservoirs were collected for Hg analysis. High concentrations of Hg were observed in muscle tissue in all species, with the highest concentrations observed in the predatory fish (Table 4-3). The relative Hg concentrations in different fish species were influenced by feeding habits with the following order: piscivorous species (*Silurus glanis, Esox lucius*) > benthophagous species with substantial portions of fish in their diet (*Perca fluviatilis*) > typical benthophagous species (*Abramis brama, Blicca bjoerkna, Carassius auratus*) > plankton-ophagous species (*Rutilis rutilis, Scardinius erthrophthalmus*). Mercury concentrations in muscle tissue were positively correlated with age, weight, and length in roach (*Rutilis rutilis*), bream (*Abramis brama*), and perch (*Perca fluviatilis*).

Table 4-3. Mercury Concentration in Muscle Tissues of Fish from Rivers and Reservoirs in the Czech Republic.

Species	N	Hg concentration (μg/g)[a]
Relatively contaminated sites		
Silurus glanis	2	0.771
Esox lucius	2	0.392
Perca fluviatilis	6	0.352 (0.032)
Abramis brama	14	0.218 (0.029)
Carassius auratus	14	0.219 (0.015)
Blicca bjoerkna	7	0.170 (0.020)
Rutilus rutilis	16	0.170 (0.021)
Leuciscus cephalus	1	0.083
Scardinius erythrophthalnus	4	0.059 (0.005)
Relatively unpolluted sites		
Perca fluviatilis	14	0.288 (0.288)
Blicca bjoerkna	21	0.185 (0.026)
Carassius auratus	4	0.153 (0.025)
Abramis brama	26	0.140 (0.016)
Rutilus rutilis	16	0.101 (0.012)

Source: Svobodová *et al.* 1999.
[a]Estimates of geometric mean with standard errors (in parenthesis).

Transfer rates of 41% and 92%, respectively, were observed for rainbow trout alevins fed either inorganic Hg or MeHg contaminated *Daphnia* for 10 days. Only results for MeHg were reported for *Gambusia affinis*. After 30 days of exposure, striking differences in bioconcentration factor for the direct route and the global route (direct + trophic) were apparent, as shown in Table 4-4.

Table 4-4. Differences in Bioconcentration Factor by Route.

Route	Water concentration MeHg	Prey	BCF
Direct	1 μg /L	–	2,500
Global	1 μg /L	50 daphnia/fish/day	27,000

4.1.4 Abiotic Factors

Previous studies on the influence of water quality and geochemical factors on MeHg bioaccumulation (with or without the use of ultraclean techniques) have implicated many of the factors and parameters likely to be important in the bioaccumulation of MeHg and its movement through the food chain. For example, total Hg contamination in fish [which is likely to be present as MeHg in near-quantitative amounts because 95% to 99% of the total mercury burden in fish is in the MeHg form (Grieb *et al.* 1990)] is thought to be a negative function of lake productivity and/or biomass because of biomass dilution effects (Hakanson 1980; Richman, Wren, and Stokes 1988). Thus, fish Hg accumulation generally declines as productivity and biomass increase. As a result, this may necessitate accounting for factors that influence lake biomass or productivity, such as lake size and depth (Rawson 1952; Ryder 1965), lake water chemistry (Ryder 1965; Conroy and Keller 1976), or local geochemistry (Conroy and Keller 1976; Ryder 1964). These factors—lake morphometry, water chemistry, and geochemistry—may also influence fish Hg accumulation in other ways. A number of water chemistry parameters have been linked directly to fish Hg accumulation (discussed below), while watershed geochemistry and lake morphometry secondarily influence lake water chemistry (Schindler 1971; Rasmussen, Godbout, and Schallenberg 1989; Shilts and Kettles 1989). Reviewed below are the lake chemistry factors that have been linked to Hg bioaccumulation in fish and also the physical, morphometric, and geological factors that will secondarily influence these lake water chemistry variables.

4.1.4.1 Water Quality Factors

pH has been the most studied water quality variable for its influence on Hg levels in fish, with almost all investigations having demonstrated higher fish MeHg contamination in fish of lakes with lower pH. The most likely explanation for this effect is the increased net rate of MeHg production at lower pH (Richman, Wren, and Stokes 1988, and reviews by Gilmour and Henry 1991; Winfrey and Rudd 1990; Beijer and Jernelov 1979).

Scheuhammer and Graham (1999) compared the relative rates of Hg accumulation in fish muscle from two Ontario lakes that were similar in all characteristics except that one, Blue Chalk Lake, was approximately neutral (pH 6.9) while the other, Plastic Lake, was acidic (pH 5.6). Sediment Hg concentrations were at or below the analytical detection limit of 5 ng/g dw in both lakes. Pumpkinseed (*Lepomis gibbosus*), creek chub (*Semotilus atromaculatus*), brown bullhead (*Ictalarus nebulosus*), and crayfish (*Orconectes virilis*) were collected from Blue Chalk Lake. Top predator species were apparently absent from both lakes. Pumpkinseed, rock bass (*Amploplites rupestris*), creek chub, and yellow perch (*Perca flavensceus*) were collected from Plastic Lake. Mean total Hg concentrations for both lakes are shown in Table 4-5. Two species were found in each lake, creek chub and pumpkinseed. Regression analysis showed that Hg concentrations were not related to length in either lake for creek chub, and t-tests showed no difference in mean Hg concentration between the two lakes. For pumpkinseed, however, a statistically significant relationship between length and Hg concentrations was observed, and the slope of the relationship was 3 to 4 fold greater for the acidic Plastic Lake. These results suggest that pH is an important parameter controlling availability and bioaccumulation of Hg in lakes that were otherwise comparable. The results also suggest that environmental conditions may be more important than the concentrations of Hg in the sediments, as in both lakes the concentration of Hg in the sediment was below 5 ng/g dw.

Table 4-5. Concentrations of Mercury in Dorsal Muscle of Fish or Crayfish from Blue Chalk Lake (pH 6.9) and Plastic Lake (pH 5.6).

Organism	Blue Chalk Lake (μg/g ww)[a]	Plastic Lake (μg/g ww)
Brown bullhead	0.11	–
Creek chub	0.08	0.10
Rock bass	–	0.22
Pumpkinseed	0.04	0.13
Yellow perch	–	0.13
Crayfish	0.04	–

Source: Scheuhammer and Graham 1999.
[a]Converted to wet weight basis using the relationship wet mass concentration = 0.22 (dry mass concentration) from Snodgrass *et al.* (2000).

Alkalinity, acid neutralizing capacity (ANC) [which is equivalent to alkalinity for most lakes (Dillon, Yan, and Harvey 1984)], as well as measures of water hardness, are directly related to pH. Thus, as pH decreases, parameters such as alkalinity and ANC decrease (Neary *et al.* 1990; Jefferies *et al.* 1986; Dillon, Yan, and Harvey 1984; Eilers *et al.* 1983). Greater Hg contamination in fish is also associated with lakes having lower alkalinity (Scheider, Jefferies, and Dillon 1979). There are two proposed mechanisms for this relationship. First, lower alkalinity may reflect lower pH and subsequent increased net Hg biomethylation. Alternatively, calcium, as a major component of alkalinity and ANC (as $CaCO_3$) (Dillon, Yan, and Harvery 1984; NRCC 1981) may regulate the permeability of fish gills to MeHg. This would explain the enhanced Hg uptake by rainbow trout (*Oncorhynchus mykiss*) that was found in soft water compared to hard water (Rodgers and Beamish 1983).

Sulfate concentration in lake water may also be associated with increased Hg biomethylation and thereby increased fish Hg contamination (Gilmour, Henry, and Mitchell 1992). As previously stated, anaerobic bacteria with sulfate-reducing capability can co-metabolize inorganic Hg to MeHg in lake sediments. Both bacterial metabolism and Hg methylation increase as sulfate concentration increases in the overlying water (Gilmour, Henry, and Mitchell 1992).

Aluminum concentration is also expected to increase with increasing sulfate deposition (NRCC 1981). It appears that sulfate deposition affects lake water concentrations of DOC. Changes in lake DOC have been reported after experimental acidification and liming of lakes (Schindler *et al.* 1992; Molot, Dillon, and Booth 1990; Bukaveckas and Driscoll 1991). Neary *et al.* (1990) showed that average DOC was lower in lakes with higher sulfate deposition, the relationship being affected by lake size. Schindler *et al.* (1992) and De Haan (1992) both postulated that lower DOC in lakes receiving acidic deposition is caused in part by reduced DOC levels in acidified runoff, due to the pH dependence of DOC solubility in runoff.

There is evidence to suggest that selenium may have some effect on the bioaccumulation of organic forms of Hg. Selenium has been shown to decrease the bioaccumulation of MeHg from sediment by oligochaete worms under experimental conditions (Nuutinen and Kukkonen

1998). In this study, a high sediment organic carbon content was also linked to a reduction in MeHg bioaccumulation. Additionally, Southworth, Peterson, and Ryon (2000) looked for changes in bioaccumulation of Hg in largemouth bass in Rogers Quarry after discharge of selenium-rich fly ash into the quarry was suspended. Rogers Quarry in Anderson County, Tennessee, had been used as a settling pond for fly ash from a coal-fired power plant for more than 20 years, with consequent selenium concentrations of 20–30 µg/L. After discharges were stopped, the selenium concentration fell to 2–5 µg/L within one year and fell below the analytical detection limit after two years. Water chemistry in the quarry was characterized by moderate alkalinity (2–3 meq/L) and a pH range of 7–9. In the eight years following cessation of fly ash discharge, the muscle Hg concentrations in largemouth bass rose from 0.02 to 0.73 µg/g ww (0.63 µg/g adjusted for covariance with weight). Although the selenium concentration in water fell rapidly, the selenium concentrations in fish tissue fell more slowly and leveled off at a concentration range of 1–1.5 µg/g, which is approximately double the background level in this area. Thus, Southworth, Peterson, and Ryon (2000) suggested that the effect of selenium is not internal to the fish, but may occur by suppressing MeHg production in sediments.

Hg contamination of fish from drainage lakes (lakes recharged by surface runoff) has a positive association with the organic carbon content of water (DOC, total organic carbon, color) (Wren *et al.* 1991; McMurtry *et al.* 1989). Both inorganic Hg and MeHg are transported to drainage lakes as a complex with organic carbon (Lee and Iverfeldt 1991; Mierle and Ingram 1991; Richardson and Currie 1995). If the flow of DOC from the watershed were reduced, the bioavailability of MeHg to fish may be reduced. Also, DOC may play an important role in the abiotic methylation of inorganic Hg (Weber 1993). Reduced lake DOC would therefore result in a reduced rate of abiotic Hg methylation in the water column and in lower relative Hg contamination in fish. Both of these proposed mechanisms would lead to a positive association between lake DOC and fish Hg levels. However, DOC has been shown to inhibit the biomethylation of inorganic Hg in controlled experiments with no constant influx of new Hg (Miskimmin, Rudd, and Kelly 1992). Thus, in seepage lakes that are not inundated with DOC-bound Hg from the watershed, lower levels of DOC in the water column may lead to an increase in the relative rate of Hg biomethylation with subsequent increased bioavailability for uptake by fish. This would explain the observed negative association between fish Hg and lake DOC levels in seepage lakes (Wiener *et al.* 1990; Cope, Wiener, and Rada 1990; Grieb *et al.* 1990).

Among more recent reports that have highlighted the influence of DOC [added as aquatic humus (AH)] and alkalinity on the uptake of MeHg by *D. magna* was the *in vitro* study of Monson and Brezonik (1999). Alkalinity (the degree of water hardness) had no effect on the uptake of MeHg by *D. magna* in 24- or 48-hour experiments. The pHs of both "soft" and "hard" culture media were adjusted to 6.5 to eliminate pH effects, the hard water medium reflecting the composition of a lake in calcareous soil and the soft water medium of an acid-sensitive lake.

In the same study, DOC present as measured additions of AH at concentrations of either 3 or 10 mg/L simulated low and high DOC softwater lakes (Monson and Brezonik 1999). Tissue total Hg increased in proportion to the DOC content of the incubation medium up to a point. However, at 10 mg/L AH, the MeHg content of tissue was lower than those of control incubations (without DOC) but with the same source of MeHg. Demethylation was ruled out as a reason for the lower concentration of MeHg in the presence of high AH. It was speculated that high AH/DOC may complex MeHg, reducing the bioavailability of the compound but, in contrast, may make available other forms of Hg.

Snodgrass *et al.* (2000) assessed the relative roles of trophic level, wetland morphology, hydrology, and water chemistry in relation to Hg concentrations in whole fish in nine southeastern depression wetlands. These wetlands are seepage lakes with no inflow or outflow (Carolina bays). In these depression wetlands, internal variables such as pH, dissolved organic matter, and sulfate levels are the predominant factors controlling Hg bioaccumulation. Maximum water depth, wetland area, and drying score (a relative measure of water level fluctuation) were determined to quantify wetland morphology. Three fish species were monitored: lake chubsucker (*Erimyzon sucetta*), which feeds on benthic detritus; mud sunfish (*Acantharchus pomotis*), which feeds on macroinvertebrates during early life stages and later shifts to a diet that includes small fishes; and redfin pickerel (*Esox americanus americanus*), which feeds on macroinvertebrates during early life stages and later shifts to a diet of shrimp, crayfish, and fish. Mercury concentrations ranged from 0.002–0.39 µg/g ww for lake chubsucker, 0.029–0.34 µg/g ww for mud sunfish, and 0.002–0.42 µg/g ww for redfin pickerel. Positive correlations between fish mass and Hg concentrations were observed for mud sunfish and redfin pickerel, but not for lake chubsucker. There were few relationships observed between the Hg content and the water chemistry variables measured. The lakes were acidic with a narrow pH range (4.5 to 5.5). In addition, the sulfate levels observed were below levels that affect methylation rates. The strongest relationships observed both by regression analysis and pathway analysis indicated negative effects of maximum water depth and hydroperiod. These results suggest that the effects of wetting and drying of the sediments in these wetland systems influence Hg bioavailability.

Hanten *et al.* (1998) examined the relationships between Hg concentrations in largemouth bass and physical and chemical characteristics in 46 lakes from five ecological regions of Connecticut. Most of the lakes (33) were drainage lakes and the rest (13) were impoundment lakes. Measured chemical characteristics included alkalinity; calcium; magnesium; DOC; total, dissolved and particulate phosphorus; and nitrogen. Physical attributes of lakes evaluated were maximum depth, mean depth, surface area, volume, watershed area, watershed to surface area ratio, retention time, shoreline development index, and elevation. Fish greater than 30 cm were targeted, and regressions were used to predict the expected Hg concentration in muscle tissue at a standardized length of 35.6 cm. Mean Hg concentrations in largemouth bass muscle ranged from 0.244 to 1.084 µg/g ww. The mean predicted bass Hg concentration was 20% higher for impoundments (0.514 µg/g) than for drainage lakes (0.410 µg/g). Principal component analysis for hardness and alkalinity, hydrology, productivity, lake size, and elevation accounted for 83% of the variation in the entire dataset of lake variables. Stepwise multiple regression showed that log particulate P and log mean depth accounted for 51% of the variation in predicted fish Hg concentration in drainage lakes. In impoundment lakes, log retention time was the only variable retained in stepwise multiple regression, accounting for 57% of the variation.

4.1.4.2 Geochemical Factors

Lake water chemistry is influenced by local geochemistry, particularly watershed buffering capacity. Lake water chemistry, including pH and alkalinity, is known to vary with the chemical nature of both watershed soils and bedrock (Ryder 1964; Conroy and Keller 1976; Kettles, Shilts, and Coker 1991; Shilts and Kettles 1989). The impact of acidic precipitation on lake water chemistry will be offset by a highly buffered watershed compared to a poorly buffered one (NRCC 1981). The degree of leaching of [Al] by acidic deposition into lakes is also expected to be greater in watersheds with lower buffering capacity (NRCC 1981; Cronan and Scholfield 1979).

Lake water chemistry is also dependent on lake and watershed morphometry. DOC depends on lake and watershed morphometry because drainage basin runoff is the primary source of DOC to lakes (Schindler *et al.* 1992; Rasmussen, Godbout, and Schallenberg 1989). Alkalinity, pH, [Al] and [SO_4^{2-}] are also expected to vary with morphometric factors that influence lake retention times and flushing rates, dilution factors, and chemical loading rates from the drainage basin (Schindler 1971).

4.1.5 Interaction Factors

Fluctuations in water chemistry variables do not happen independently of one another. For example, pH and alkalinity are strongly interrelated (Eilers *et al.* 1983). Therefore, a significant change in one implies some change in the other. These interactions can significantly influence or control fish Hg accumulation, as demonstrated by Richardson and Currie (1995). Several of these potential interactions and the impact on Hg concentrations are discussed below.

In lakes receiving sulfate (acidic) deposition, levels of SO_4^{2-} increase as the rate of deposition and degree of lake acidification increase (Sullivan *et al.* 1988; Neary *et al.* 1990; Howells 1990). Therefore, declining pH and declining alkalinity will be associated with a concomitant increase in SO_4^{2-} in these lakes.

The solubility of DOC is pH-dependent (Thurman 1985), declining at lower pH; and because DOC also has some acid neutralizing capacity (Schindler *et al.* 1992), the concentration of DOC in water will decline with increasing acidification. Overall, this will be slower in lakes with higher relative alkalinity, because of the buffering effects of alkalinity and competition between DOC and alkalinity for ANC. Finally, DOC directly influences pH at relatively high DOC levels (dystrophic lakes) (Thurman 1985).

It has been hypothesized that aluminum concentration [Al] may control DOC levels in acidified lakes through the formation and precipitation of DOC-aluminum complexes (Effler, Schafran, and Driscoll 1985). However, Schindler *et al.* (1992) suggested that the postulated link between DOC and [Al] may be incorrect. Instead, it may be that [Al] simply reflects the concentration of organic ligands (such as humic and fulvic acids) with which it tends to complex in natural waters (Howells 1990).

4.2 Phytoplankton and Zooplankton

Phytoplankton constitute the first link of the trophic chain in the lacustrine environment. Boudou and Ribeyre (1997a,b), drawing largely on the results of their own earlier laboratory investigations, set up an experimental four-link trophic chain to compare the transfer of inorganic Hg and MeHg between consumers for both direct and trophic routes. The direct route was defined as the direct uptake of contaminants from water by an aquatic organism (absorption) and the trophic route as the uptake of contaminants from prey. The experimental trophic chain comprised phytoplanktonic algae (*Chlorella vulgaris*), zooplanktonic cladocerans (*Daphnia magna*), planktivorous fish (*Gambusia affinis* and *Onchorynchus mykiss alevins*), and carnivorous fish (adult *Onchorynchus mykiss*).

At 18°C, a rapid, direct uptake of both inorganic Hg and MeHg was demonstrated for the primary producer, *C. vulgaris*, with little difference between the bioconcentration factors for either species of Hg. Although the rates of inorganic Hg and MeHg uptake were followed for 24 hours, concentrations of both mercury species reached approximately maximal and close to equal values in the algae after eight hours.

The first trophic transfer in the chain, *C. vulgaris* to the herbivorous species *D. magna*, showed distinct differences in the transfer of inorganic Hg and MeHg. At an ambient temperature of 18°C, the transfer rate for inorganic Hg was 6%, and for MeHg 58%.

These studies suggest that trophic transfer between algae and herbivorous zooplankton plays a key role in the biomagnification of MeHg. By contrast, inorganic Hg may be bound to the algal cell wall, with eventual depuration by herbivorous species, thereby losing potential for biomagnification. Conversely, MeHg contained within the cytoplasm of the algae is efficiently assimilated by the herbivore, thus identifying in the first trophic transfer the mechanism by which biomagnification occurs.

4.3 Macroinvertebrates

Aquatic macroinvertebrates, or macroinvertebrates that have benthic larval stages, may be an important source of dietary Hg for some terrestrial species, as well as higher trophic level aquatic predators. For example, the Asiatic clam (*Corbicula fluminea*) is a prey item for bream (*Sparidae*), tench (*Tinca tinca*), crayfish (*Astacus astacus*), muskrat (*Ondatra zibethicus*), and coypu (*Myocator coypus*). Macroinvertebrates represent additional links in the food web between primary producers and top-tier receptors. Therefore, it is of value to have an estimate of the bioaccumulation of MeHg in macroinvertebrates and the efficiency of trophic transfer between them and their predators.

Inza *et al.* (1997) investigated the bioaccumulation of cadmium, inorganic Hg, and MeHg in the Asiatic clam in relation to contaminant concentration in sediment and the water column. Asiatic clam is a sediment-dwelling, filter-feeding mollusc that is widely regarded as a useful bioindicator of metallic contamination. Filter-feeding molluscs such as Asiatic clam may filter up to 20 L of water in a day and are vulnerable to contaminants present in the water column. In the experiment, exposure units (EUs), 12- × 12- × 30-cm glass tanks were set up to study uptake from either water or spiked sediment. Sufficient numbers of EUs were employed to study the bioaccumulation effects of five concentrations of each contaminant from either water or sediment. Asiatic clams were exposed to the contaminant source for 14 days.

The bioaccumulation of metals from the water column was much greater than from sediment, although the concentration of contaminants in sediment was up to 2000 times higher than water EUs. MeHg bioaccumulated more readily from either the water column or sediment than did cadmium or inorganic Hg. After a 14-day water column exposure to 1.5 µg MeHg/L, the soft body of the clams contained more than 6000 ng/g. Bioconcentration factors estimated by the authors are presented in Table 4-6.

Table 4-6. Bioconcentration Factors in *Corbicula fluminea* –14-Day Exposure.

Contaminant	Concentration in Water Column (μg/L)	Bioconcentration Factor
MeHg	1.5	4000
Inorganic Hg	1.5	300-500
Cadmium	1.5	150

Source: Inza *et al.* 1997.

Individual organs of the bivalves were assessed for contaminant concentration. High concentrations of MeHg were present in all organs with greater concentrations present in the gills and foot. Inorganic Hg accumulated chiefly in the visceral mass. A bioconcentration factor greater than 7000 was estimated for MeHg in the gills of the Asiatic clam. The affinity of MeHg for the muscular foot of the clam shows some similarity to the bioaccumulation observed in the skeletal muscles of trout and the adductor muscle of the oyster.

The efficiency of trophic transfer of MeHg between the Asiatic clam and the crayfish has been investigated by Simon, Ribeyre, and Boudou (2000). The bioaccumulation of MeHg and cadmium was assessed in crayfish after exposure (via ingestion) to the soft bodies of contaminated clams. Crayfish were exposed each day to the opened clams, which were contaminated with MeHg or cadmium, for periods of either 5, 10, or 15 days. Table 4-7 summarizes the results.

Table 4-7. Trophic Transfer from *Corbicula fluminea* to *Astacus astacus*.

Contaminant	Exposure (%)		
	5 Day	10 Day	15 Day
MeHg	14.5	18.5	15.7
Cadmium	<1	6	4.4

Source: Simon, Ribeyre, and Boudou 2000.

MeHg was preferentially accumulated in the green gland and the tail muscle of the crayfish, with the highest levels in the muscle. The accumulation of MeHg in the tail muscle of the crayfish is a risk to predators, who may only consume the muscle.

The effects of temperature and pH on the rate of depuration of MeHg in burrowing mayfly nymphs (*Hexageniarigida*) were investigated by Odin, Ribeyre, and Boudou (1997). Citing earlier work, they reported an increase in the rate of Hg bioaccumulation by nymphs from sediment as temperature increased from 10°C to 26°C. When exposed to Hg from the water column, the insects' Hg concentration increased between 10°C and 18°C but decreased between 18°C and 26°C. Acidification of the water column to pH 5.0 increased the bioaccumulation of inorganic Hg, but decreased bioaccumulation of MeHg.

4.4 Fish

Recent research findings indicate that several variables are important in evaluating bioaccumulation of Hg in piscivorous fish, including the size and sex of the fish, trophic level, and, in general, the interplay of water quality and geochemical factors (Appendix C). Factors related to the hydrology of the water body that affect wetting and drying of sediments are also important in determining Hg bioaccumulation in piscivorous fish.

Positive correlations between Hg concentrations in muscle tissue and the age, length, and weight of piscivorous fish have been reported by a number of investigators in several fish species (Munn and Short 1997; Neumann and Ward 1999; Foster, Drake, and DiDomenico 2000). Henry *et al.* (1998), however, found that Hg concentrations in smallmouth bass were not linearly correlated with weight. Possible explanations for this observation include the fact that the fish in the study were not separated based on sex and availability of Hg in Fumee Lake may have been limited due to the alkaline (pH 8.4) conditions. Foster, Drake, and DiDomenico (2000) reported that Hg concentrations were positively correlated with age, length, and weight in male large-mouth bass, but not in females. The observation that Hg accumulation increases with age, length, and weight of fish should be considered when establishing sampling protocols. For example, when evaluating potential human health hazards, the size of the fish sampled should be selected to reflect the fact that only fish that exceed legal size limits can be kept for consumption. For evaluating ecological risks, the typical sizes of fish taken as prey differ among bird species. For example, loons are limited to smaller fish, while osprey prefer larger fish.

Accumulation of Hg increases under low pH conditions (Rose *et al.* 1999; Scheuhammer and Graham 1999). In a survey of 24 lakes in Massachusetts with low levels of Hg contamination, Rose *et al.* (1999) found that low water pH was correlated with elevated Hg concentrations in both brown bullhead and yellow perch, but not in largemouth bass. Scheuhammer and Graham (1999) observed that the slope of the relationship of Hg concentration to length in pumpkinseed sunfish was three to four fold greater for the acidic Plastic Lake (pH 5.6) than for the less acidic Blue Chalk Lake (pH 6.9). Thus, models for predicting Hg bioaccumulation may need to include an adjustment for water pH; low pH is thought to increase Hg availability by enhancing Hg methylation.

Kim and Burggraaf (1999) studied bioaccumulation of Hg in rainbow trout in five geothermal lakes in New Zealand with comparatively high Hg levels. Using clean techniques, they measured concentrations of Hg and MeHg in water, prey, and in trout. The average MeHg concentration in trout muscle varied linearly with the total Hg concentration in the water ($n = 5$, $r = 0.99$, $p = 0.01$), but no relationship with MeHg in the water was apparent. The greatest bioaccumulation occurred between water and the zooplankton in these lakes (>50,000 times), while MeHg was accumulated approximately 4.1 times between trout and smelt (*Retropinna retropinna*). In addition, statistically significant linear regressions were observed between rainbow trout MeHg concentrations and concentrations in other prey species such as bullies (*Gobiomorphus cotidianus*) and koura (*Paranephrops planifrons*).

Castilhos and Bidone (2000) studied Hg bioaccumulation in an uncontaminated fluvial lacustrine system in comparison to a river area with elevated Hg contamination due to gold mining. As shown in Table 4-8, total Hg concentrations in muscle tissues were approximately 10-fold higher in piscivorous fish compared to detritivorous, herbivorous, or omnivorous fish. This observation applied to both the Hg-contaminated and background areas. Because concentrations of Hg in the water were not determined, it was impossible to calculate bioconcentration factors from these data, although the authors estimated bioconcentration factors of 10,000 times for the noncarnivorous fish and 100,000 times for the carnivorous fish based on other published data from the region.

Table 4-8. Total Mercury Concentrations in Muscle Tissues of Carnivorous and Non-carnivorous Fish From Background and Contaminated River Systems in Amazonia, Brazil.

Feeding Habit	Mean Hg concentration (µg/kg) (mean ± SD)	
	Background	Contaminated
Carnivorous		
Piscivorous	236 ± 189	448 ± 195
Omnivorous	206 ± 104	381 ± 281
Noncarnivorous		
Detritivorous	26 ± 20	90 ± 67
Herbivorous	30 ± 40	49 ± 41
Omnivorous	36 ± 13	72 ± 30
Planktivorous	149 ± 60	–

Source: Castilhos and Bidone 2000.

These studies, however, do not allow a detailed evaluation of the bioaccumulation of Hg between these trophic levels, because clear predator-prey relationships were not demonstrated. That is, the investigators did not examine the stomach contents of the predatory fish to establish prey species and size preferences.

The hydrology of the water body also has an important effect on Hg bioaccumulation in piscivorous fish (Snodgrass *et al.* 2000; Hanten *et al.* 1998). In a survey of Hg bioaccumulation in southeastern depression wetlands, Snodgrass *et al.* (2000) found negative effects of maximum water depth and longer hydroperiod on Hg concentrations in fish. These results suggest that the effects of wetting and drying of the sediments in these wetland systems control Hg bioavailability. Hg accumulation increases in fish when streams are impounded to form reservoirs. This observation was supported by the results of Hanten *et al.* (1998) who surveyed Hg concentrations in fish from 46 lakes in Connecticut. They found that the mean size-adjusted largemouth bass Hg concentration was 20% higher for impoundments than for drainage lakes. In developing site-specific estimates of Hg bioaccumulation, the potential for enhanced Hg bioavailability due to wetting and drying of sediments may need to be considered in situations where water levels are expected to fluctuate throughout the year.

4.5 Seasonal Impacts of Methylmercury Bioaccumulation to Phytoplankton, Zooplankton, and Fish

Herrin *et al.* (1998) described the seasonal changes in the concentration of total Hg and MeHg in Devil's Lake, Wisconsin. This two-year study (1994 and 1995) assessed the buildup and storage of MeHg in the hypolimnion during the spring and summer months. A relationship between DO and MeHg was observed over the life of the study. Increasing concentrations of MeHg were associated with decreasing DO concentration in the hypolimnion, although a causal relationship could not be established. Nonetheless, the inverse relationship between MeHg and DO was both spatial and temporal. During fall destratification, the stored MeHg became available to the lake biota as the hypolimnion eroded and remained available for a short time after complete turnover.

Paralleling the seasonal changes seen in the filterable MeHg content of the hypolimnion, seasonal dynamics were apparent for MeHg associated with particulate fractions. Particulates were classified by size into edible (2.2–35 µm) and inedible (35–243 µm) fractions. The MeHg content of both fractions tripled during September and October when the lake's strata eroded. However, the MeHg concentration of the edible fraction fell rapidly as the particles were consumed by zooplankton. The initial increase of MeHg in both fractions implies MeHg is rapidly taken up by particulates as mixing takes place. This uptake of MeHg by passive diffusion is considerable, as the edible particulate fraction was found to be composed of phytoplankton, generally *Chromonas* and *Cryptomonas sp*. This provided a direct measure of the mechanism by which MeHg enters the lacustrine food web from the anoxic waters of the hypolimnion.

Samples of water, particulate matter (edible and inedible), and *Daphnia* were collected for analysis using ultraclean techniques. Yearling mimic shiners (*Notropis volucellus*) and bluegills (*Lepomis macrochirus*) were also collected for Hg analysis. Results of these analyses are presented in Table 4-9.

Table 4-9. Devil's Lake, Wisconsin: Mercury Concentrations in the Water Column and at Different Trophic Levels of the Aquatic Food Chain.

Year	Water Concentration[1]		MeHg in Edible Particles[2]	MeHg in *Daphnia*	Mercury in Bluegill[3] (µg/kg)		Mercury in Shiners[3] (µg/kg)	
	Hypolimnion (ng/L)		ng/g (dw)	ng/g (dw)	dw (ww)	dw (ww)	dw (ww)	dw (ww)
	Me Hg	Total Hg	MeHg	MeHg	MeHg	Total Hg	MeHg	Total Hg
1993	1.2	3.0	–	–	837 (185)	814 (180)	–	–
1994	0.52	2.0	9.6	186	509 (115)	575 (129)	483 (115)	506 (120)
1995	0.07	1.2	9.1	100	265 (58)	324 (74)	287 (70)	336 (82)

Source: Herrin *et al.* 1998.

[1]Based on running daily average concentrations, 7–12 September and 3–11 October for 1994 and 1995. September 30, 1993 was the only sampling day for that year.

[2]Based on running daily average, 12 September and 15 November.

[3]Concentrations for fish born in the listed year and sampled the following spring.

Bioconcentration factors were not calculated, as all MeHg concentrations in whole water from the epilimnion were below detection limits. Using the equation given below, the authors calculated biomagnification factors (BMF). These results are presented in Table 4-10.

$$BMF = \frac{[mass\,Hg / dry\,mass\,predator]}{[mass\,Hg / dry\,mass\,prey]} \qquad (4.5\text{-}1)$$

Table 4-10. Devil's Lake, Wisconsin: Biomagnification Factors for Methylmercury.

Year	Edible Particles to *Daphnia*	*Daphnia* to Mimic Shiners	*Daphnia* to Bluegill
1994	19	2.6	2.7
1995	11	2.9	2.7

Source: Herrin *et al.* 1998.

The seasonal patterns of mercury species in water and zooplankton were also investigated in 12 softwater lakes in Minnesota (Monson and Brezonik 1998). Two years of measurements were made to assess seasonal effects on labile (unextracted) MeHg and total (extractable) MeHg in the water column and total Hg and MeHg in zooplankton. Total MeHg was solvent extracted and back extracted into Hg-free water before analysis. This form of Hg was assumed to include free MeHg dissolved in water and particle-bound MeHg. Labile MeHg was assumed to include free MeHg dissolved in water and loosely bound MeHg and was not solvent extracted prior to analysis. Both total MeHg and labile MeHg were analyzed to determine if there were any differences in their bioavailability. Clean sampling and analytical protocols were used throughout the project. In addition to the seasonal dimension to this experiment, water quality parameters were also measured, including acid neutralizing capacity (ANC), pH, DOC, color, sulfate, chlorophyll, and total phosphate. No statistically significant correlation was demonstrated between any water quality parameter and either total Hg or MeHg concentration in plankton.

Seasonal changes were seen in water mercury species. Water and plankton data for the two-year time frame of the study were combined from all 12 lakes, and the mean values (+/– 1 S.E.) compared for the three seasons: spring, summer and fall. MeHg declined throughout the growing season, but the average decline was greater for total MeHg than for labile MeHg. Total Hg declined in water from spring until summer but was observed to rise in the fall. The authors suggest that the high spring and fall total Hg levels may reflect mixing. Mercury concentrations in plankton did not follow the seasonal pattern seen in water and were lowest in spring. The greatest concentrations were in fall, four times higher than those in spring. The mean percent of MeHg in zooplankton rose from 20% in spring to 52% in the fall.

Results from this investigation indicate a large disparity in the BAFs calculated for fish and plankton and suggest that plankton concentrations of MeHg are not in equilibrium with water concentrations. One order of magnitude separated BAFs for plankton and fish (young-of-the-year bluegill). For total Hg there was 4.5 orders of magnitude difference between water and plankton, and for MeHg, 5.5 orders of magnitude between water and plankton. Log BAF for MeHg in plankton was either 4.90 or 5.43, depending on the form of MeHg in water used for the calculation (labile or total). However, the lack of equilibrium between water MeHg and plankton is indicated by the increasing MeHg concentration of plankton at the time water MeHg is declining. This indicates that BAFs for MeHg in plankton are variable and increase from spring to fall. The BAF for labile MeHg in plankton increases particularly rapidly in fall. In interpreting their results, the authors suggested that labile MeHg may be a useful indicator of bioavailable MeHg, because it follows the seasonal pattern of total MeHg.

Foster, Drake, and DiDomenico (2000) followed seasonal changes in Hg concentrations in muscle, liver, and gonads of largemouth bass in Dorena Reservoir, Oregon. Park and Curtis (1997) reported seasonal variations in Hg concentration in largemouth bass muscle tissue. However, Foster, Drake, and DiDomenico (2000) found no difference in muscle Hg concentration from one month to another. Average muscle Hg concentrations ranged from 0.33 to 500 μg/kg ww for average fish with ages ranging from 4.8 to 6.0 years. Significant correlations between weight, length, and age and muscle Hg concentrations were observed in males but not in females. Liver and gonad Hg concentrations were also correlated with weight in males but not in females. Stomach contents of largemouth bass collected in the spring were mostly mayfly larvae, but in the summer the contents were mostly fish. Only minor seasonal variation in liver and gonad Hg concentrations in males were observed. These data point to the possibility that seasonal factors likely play only a minor role in affecting Hg bioaccumulation at the higher trophic levels.

Watras *et al.* (1998), in a study of 15 north Wisconsin lakes, used field data, equilibrium aqueous speciation modeling, and statistical data to re-examine assumptions on the uptake, distribution, and fate of total Hg and MeHg in pelagic food webs. Field data for each of the 15 lakes included the concentration of total Hg and MeHg in surface water, sediment, microseston, zooplankton, and small fish. The water quality characteristics determined for each lake included the analysis of DOC, pH, dissolved inorganic carbon (DIC), small particulate matter (SPM), chlorophyll a, conductivity, calcium, magnesium, sodium, iron, manganese, sulfate, and chloride. In these lakes, the microseston primarily comprised phytoplankton, bacterioplankton, and cellular debris. Zooplankton were sorted and the results evaluated by genera. The fish species examined were yellow perch (*Perca flavescens*) and golden shiners (*Notemigonus crysoleucas*).

The concentrations of both MeHg and total Hg in microseston were negatively correlated with DOC. Seston-water partition coefficients were twice as high for MeHg as for total Hg, an observation possibly reflecting stronger binding of total Hg to DOC. A slight increase in MeHg concentration in seston was seen as pH decreased, in parallel to a slight increase in dissolved MeHg. In microseston, log BAF values for inorganic Hg ranged from 4.7 to 5.9 and for MeHg from 4.8 to 6.2. BAFs for higher trophic levels increased only by a factor of 2 to 4, indicating that the first step in the bioaccumulation process may be the most important. The bioaccumulation of MeHg or inorganic Hg by microseston was independent of chloride concentration, although this parameter varied by a factor of 30 among the 15 lakes studied. MeHgCl concentrations were calculated for each lake, but these were not found to correlate with MeHg bioaccumulation in microseston. A positive correlation was observed between MeHg in microseston and MeHgOH and $MeHg^{+}$.

Concentrations of MeHg and the percentage of MeHg were higher in zooplankton than in microseston, but the concentration of inorganic Hg was markedly lower in zooplankton than microseston. This observed biomagnification of MeHg confirmed the results of previous studies performed by the authors. MeHg levels in zooplankton were negatively correlated with pH and positively related to DOC. However, when zooplankton were considered as a group rather than as individual genera, the correlation was weak. MeHg levels in the genus *Diaptomus* showed stronger correlation with these parameters than most other genera that were evaluated.

Perch (approximate weight 30 g) were chosen as the representative species to assess the impact of Hg bioaccumulation for all 15 lakes. Mercury concentration in perch showed a strong negative correlation with pH ($r^2 = 0.72$) and a weak (though statistically significant) positive correlation ($r^2 = 0.33$) with DOC. These findings reflect the patterns observed for microseston and zooplankton, although the correlation was stronger than for the latter groups. Mean MeHg concentrations (μg/kg dw) for the 15 lakes were microseston 34, zooplankton 53, and small fish 485.

4.6 Birds and Wildlife

Interpretation of bioaccumulation of Hg from fish to piscivorous birds is complicated by isolation of a large fraction of the total body burden of Hg in the plumage and by their migratory behavior. DesGranges *et al.* (1998) found that approximately 85% of the total body burden of 40-day-old osprey chicks was present in the feathers. Scheuhammer *et al.* (1998) found that 60% to 90% of the body burden of loons could be accounted for in feathers. Mercury is tightly bound to keratin and deposited in the feathers. The blood supply to feathers is cut off after growth is complete, thus effectively isolating Hg deposited in feathers. Because feathers can be collected with minimal damage to the birds, several attempts have been made to predict Hg concentrations in other tissues from Hg concentrations in feathers. These attempts have met with mixed success. Scheuhammer *et al.* (1998) found that Hg concentrations were correlated to blood Hg concentrations in both chicks and adults in loons. In contrast, Caldwell, Arnold, and Gould (1999) reported poor correlation between Hg concentrations in feathers, blood, tissues, and eggs in double-crested cormorants. Wolfe and Norman (1998) found a strong correlation between feather Hg, and both liver and brain Hg in double-crested cormorant chicks, but the results were mixed for great blue heron chicks. Correlation between feather Hg concentration and other tissues in adult piscivores is complicated by the fact that these birds typically overwinter in Central and South America, and a new set of feathers are grown after molting in the overwintering area. Thus, poor correlation between feather and tissue or blood Hg in adults can be easily explained. The poor correlation for chicks has not been satisfactorily explained.

Based on a comparison of feather Hg concentration to prey Hg concentration, BMFs in excess of 10 can be calculated (Halbrook, Brewer, and Buehler 1999; Hughes, Ewins, and Clark 1997; DesGranges *et al.* 1998). Because a large portion of the Hg is isolated in the plumage, however, BMFs for other tissues tend to be much lower. Comparing Hg concentrations in bird muscle tissue to fish produced no evidence of biomagnification in double-crested cormorants (Hughes, Ewins, and Clark 1997). Similarly, DesGranges *et al.* (1998) reported that there appears to be no biomagnification of Hg from ingested food to blood in ospreys.

The same factors that affect bioaccumulation of Hg in fish also affect bioaccumulation in piscivorous birds. Meyer *et al.* (1998) found a significant negative correlation between lake pH and blood Hg in loon chicks in a survey of 46 lakes in Wisconsin. Scheuhammer *et al.* (1998) also found that lake pH was correlated with blood Hg concentrations in both adults and chicks across 24 Ontario lakes. Blood and feather Hg concentrations were five times higher in osprey chicks reared near a reservoir in contrast to natural lakes in the La Grande area in Ontario (DesGranges *et al.* 1998). This is not surprising as piscivorous birds receive their Hg dose from fish consumed. The commonly observed positive correlation between fish age, length, weight and Hg concentrations is also important in estimating Hg intakes by piscivorous birds. Thus, loons that consume small fish in the range of 20–50 g will receive a lower Hg dose than osprey that generally consume larger fish.

Meyer *et al.* (1998) conducted a long-term study on the impact of Hg exposure on reproduction and survival of the common loon (*Gavia immer*) at 46 lakes in northern Wisconsin. 75% (all <100 ha in size) of the lakes were seepage lakes with no inlet or outlet, while the remainder were drainage, drained, or spring-fed lakes. The study lakes were oligotrophic to mesotrophic with a range of pH from 4.9 to 9.5 and ANC of 9 to 1513 µeq/L. A significant negative linear relationship between log (chick blood Hg concentration) and lake pH was observed. Inclusion of lake ANC and lake area as predictor variables in stepwise multiple regressions further improved the prediction of chick blood Hg concentration. In addition, a significant negative linear relationship between adult male and female log (blood Hg concentration) and lake pH was observed. The addition of other lake predictor variables by stepwise multiple regression did not improve the fit of the relationship. Total Hg concentrations in eggs (0.9 ± 0.3 µg/g fw) were within the range associated with reduced hatching and survival for sensitive species. A significant negative linear relationship between chick survival to fledging (three-year total) and chick blood Hg concentration was observed. The relationship extrapolated to zero fledging success when blood Hg concentration exceeded 0.5 µg/mL whole blood. Finally, the number of adult loons returning to their breeding territories for the subsequent three-year period was independent of adult blood Hg concentration. These data do not allow calculation of a BAF for Hg in piscivorous birds, but they do suggest that the same factors that affect Hg accumulation in fish also affect birds. For example, yellow perch are the preferred prey item fed to common loon chicks, and perch Hg concentrations are negatively correlated with lake pH (Scheuhammer *et al.* 1998). Common loon chicks are fed prey items exclusively from the nesting lake, but adult loons in this area of small lakes also hunt at other nearby lakes when foraging for their own food needs. Thus, the higher correlation between chick blood Hg and lake pH would be expected.

Scheuhammer *et al.* (1998) examined the relationship between water chemistry, fish Hg concentrations, and Hg concentrations in the blood and feathers of common loons on 24 lakes in Ontario. Blood Hg levels in adult loons were approximately one order of magnitude higher than in chicks; the overall levels correlated with size. Male loons also had significantly higher blood Hg concentrations, which may have been related to ingestion of larger prey items with higher Hg concentrations. The plumage of loons can comprise 60% to 90% of the total body burden of Hg. Thus, Hg concentrations in feathers are an important factor in the Hg mass balance in loons. Blood and feather Hg concentrations were significantly correlated in both adults and chicks.

Another important aspect of the study was that Hg concentrations in muscle tissue of small, 20–50 g yellow perch, pumpkinseed sunfish, or rock bass (typical loon prey items) were also determined. Fish Hg concentrations ranged from 0.23 to 2.0 µg/g dw in fish from the 24 lakes, and 94% of the total Hg in the fish was MeHg. Multiple linear regression analysis found that DOC was positively correlated with fish Hg concentration and lake pH was negatively correlated with fish Hg concentration ($Hg_{fish} = 0.10 DOC - 0.26 pH + 2.36$; $R^2 = 0.445$; $p < 0.05$). When predator and prey data were combined, a significant positive correlation was observed between fish Hg concentration and blood Hg concentration in both adults and chicks. Lake chemistry parameters that affected fish Hg concentrations were also reflected in the Hg concentrations in blood of both loon adults and chicks.

Hughes, Ewins, and Clark (1997) investigated Hg accumulation in osprey (*Pandion haliaetus*) at three natural lakes and one reservoir in Ontario, and at two sites in New Jersey. Ospreys are large piscivorous raptors, their Hg body burdens representing a mixture of exposure on the North American breeding grounds and from overwintering areas on the major water systems in South America. By sampling eggs, feathers from chicks, and feathers from adults, Hughes, Ewins, and Clark (1997) examined the contribution of the local environment to Hg body burdens. Yellow perch, a major component of the osprey diet, were also sampled from the North American study sites as an indication of spatial variation in Hg availability in the diet. Variation in chick feather (28–35 days old) Hg concentrations among study sites more closely matched observed spatial variation in yellow perch Hg concentrations (Table 4-11). Thus, chick feather Hg concentration was considered to be a better indicator of Hg bioavailability in the breeding area than egg or adult feather Hg concentrations. The Hg biomagnification factor, calculated as the ratio of egg dw Hg concentration to yellow perch (4- to 5-year-old fish) muscle dw Hg concentration, ranged from 1.07 to 2.05 for the three Ontario lake sites (Table 4-11). Much higher biomagnification factors of 12.0–13.6 were calculated based on the ratio of chick feather dry weight Hg concentration to yellow perch Hg concentration (Table 4-11). This is an indication that isolation of Hg in feathers is a major pathway of Hg elimination in birds, as Hg is tightly bound to keratin synthesized during feather growth. In contrast to the natural lake areas in Ontario and the estuarine areas in New Jersey, eggs and chick feathers from the breeding ground around the Ogoki Reservoir, flooded in 1943, had significantly higher Hg concentrations. This observation agrees with the general observation that Hg levels in aquatic biota are generally elevated in artificially impounded reservoirs.

Table 4-11. Concentration of Mercury (µg/g dw) in the Eggs, Chick Feathers (28–35 days old), and Adult Feathers of Ospreys Collected at Three Natural Lake Areas, a Reservoir, and Two Estuary Areas.

Location	Eggs 1991–1992		Chick Feathers 1991–1994		Adult Feathers 1991		Yellow Perch (4–5 years)[b]	
Area	n	Mean[a]	n	Mean	n	Mean	n	Mean
Ontario								
St.Mary's River	8	0.64b	12	7.40a	2	28.8a	7	0.60a
Georgian Bay	17	0.78ab	13	4.56b	5	21.1ab	4	0.38b
Kawartha Lakes	22	0.44b	19	3.26bc	4	6.7b	14	0.24c
Ogoki Reservoir	8	1.40a	5	10.98a	–	–	–	–
New Jersey								
Delaware Bay	–	0.42b	14	2.14c	–	–	–	–
Atlantic coast	–	–	7	3.35bc	–	–	–	–

Source: Hughes, Ewins, and Clark 1997.
[a]Means followed by the same letter are not significantly different among areas comparison of means ($p < 0.05$).
[b]Yellow perch were collected by the Ontario Ministry of Natural Resources from nearby areas in either 1988, 1989, or 1992.

DesGranges *et al.* (1998) monitored concentrations of Hg in the tissues of osprey chicks reared near reservoirs and in chicks reared near natural lakes and rivers in northern Quebéc to evaluate biomagnification from prey to birds. Osprey are particularly susceptible to Hg accumulation because they generally consume larger fish that are often piscivorous fish species with higher Hg concentrations. The mean total quantity of Hg in 40-day old chicks was 10.5 mg from nests near the La Grande reservoirs compared to 1.6 mg from nests in natural areas near lakes and rivers in the James Bay and Hudson Bay regions. The average proportion of total Hg as MeHg was greater than 90% in eggs, stomach contents, blood, brain, and muscle, as compared to 74% in liver and 76% in kidneys. The concentration of Hg in the diet of osprey chicks was estimated using fish parts found near nests. Diet species identification and estimates of size distribution of the fish consumed were also done. Linear regression of the estimated Hg concentration in the fish consumed against chick blood Hg concentration showed a significant linear relationship. Approximately 85% of the body burden in 40-day-old chicks from both reservoirs and natural areas was found in the feathers. This indicates that eliminating Hg in feathers is an important mechanism protecting chicks against the toxic effects of Hg. Although elevated levels of Hg were observed in areas around reservoirs, no effect on the number of chicks fledged per nest was observed. An average of 1.6 young per nest were fledged in areas with the highest Hg levels, which was not significantly different from the rate of 2.0 young fledged per nest in natural habitats with lower Hg levels.

Halbrook, Brewer, and Buehler (1999) investigated concentrations of Hg in eggs, liver, muscle, and feathers of great blue heron (*Ardea herodias*) chicks from two colonies on the Oak Ridge Reservation (ORR), Oak Ridge, Tennessee, and two colonies located off the ORR. Concentrations of Hg were also determined in fish regurgitated from the crops of nestlings or collected beneath the nests. Previous industrial operations on the ORR have introduced a variety of contaminants into the surrounding environment. One great blue heron colony was located within the boundaries of the K-25 Oak Ridge Gaseous Diffusion Plant near Poplar Creek, and the second, Melton Hill colony, was located on three small islands in Melton Hill Lake within 1 km of ORR. The two reference colonies, Long Island and Looney Island, were located on the Tennessee River upstream of ORR. The collected fish samples were 66% shad (*Dorosoma sp.*), 11% a mixture of carp (*Cyprinus carpio*), catfish (*Ictalurus punctatus*), and bass (*Micropterus sp.*), and the remaining 19% could not be identified. The concentration of Hg in the fish collected from the K-25 colony averaged 0.08 µg/g dw and was significantly higher than the other three colonies (Table 4-12). Biomagnification of Hg was apparent for chick feathers compared to forage fish and ranged from 25 to 87 fold among the four colonies. Accumulation of Hg was also observed in eggs (2–14 fold), liver (3.6–13 fold) and in muscle (1.1–7 fold) in great blue herons. Although Hg bioaccumulated from forage fish to great blue heron tissues, there was no apparent effect on reproductive success, nor was.there any significant difference in the mean number of chicks fledged per nest among the four colonies. The average value of 2.6 fledglings per nest was also similar to that observed at other locations.

Table 4-12. Concentrations of Mercury (μg/g dw) in Regurgitated Fish and in Eggs, Feather, Liver, and Muscle from Great Blue Heron (Ardea herodias) Chicks Collected from Colonies On and Off the Oak Ridge Reservation (ORR).

	On ORR		Off ORR		
	K-25[ab]	Melton Hill	Long Island	Looney Island	p
Fish	0.08 ± 0.03[a]	0.02 ± 0.01[b]	0.01 ± 0.00[b]	0.02 ± 0.01[b]	0.007
Eggs	0.17 ± 0.02 (24)	0.13 ± 0.06 (9)	0.14 ± 0.02 (24)	0.12 ± 0.02 (10)	0.083
Feather	2.02 ± 0.28 (26)[a]	1.02 ± 0.13 (12)[b]	0.87 ± 0.08 (23)[b]	1.02 ± 0.16 (12)[b]	<0.001
Liver	0.29 ± 0.04 (26)[a]	0.15 ± 0.02 (12)[b]	0.12 ± 0.01 (23)[b]	0.26 ± 0.04 (12)[a]	<0.001
Muscle	0.09 ± 0.01 (25)	0.05 ± 0.01 (12)	0.07 ± 0.01 (22)	0.08 ± 0.01 (12)	0.077

Source: Halbrook, Brewer, and Buehler 1999.
[a]Average ± standard error (sample number).
[b]Means followed by the same letter are not significantly different ($p < 0.05$, Fisher's least significant difference test).

Caldwell, Arnold, and Gould (1999) examined Hg concentrations in eggs, blood, liver, muscle, and feathers of nestling double-crested cormorants (*Phalacrocorax auritus*) from the Elephant Butte and Caballo Reservoirs on the Rio Grande River in south central New Mexico. They also determined Hg concentrations in typical prey fish species, gizzard shad (*Dorosoma cepedianum*), white bass (*Morone chrysops*), and white threadfin shad (*Dorosoma petenense*). White bass feed on invertebrates and juvenile fish and were observed to have the highest concentrations of Hg (0.21 μg/g ww) among the three fish species examined. In comparison, Hg concentrations of 0.13 and 0.05 μg/g ww were determined in whitefin shad and gizzard shad, respectively. Gizzard shad feed on zooplankton as juveniles and switch to bottom-filter feeding detritivores as adults. The diet of adult whitefin shad consists primarily of diatoms and occasional fish larvae. Little difference in Hg concentrations of double-crested cormorant eggs or tissues was observed between the two reservoirs (Table 4-13). Because Caldwell, Arnold, and Gould (1999) did not report quantitative data on the composition of fish species in the double-crested cormorant diet, no conclusions can be drawn concerning bioaccumulation of Hg from fish to birds. However, based on a comparison of breast muscle tissue in the nestling cormorants to muscle tissue of white bass, there was no evidence of bioaccumulation of Hg between these trophic levels. These two reservoirs are considered to be relatively uncontaminated with reported MeHg concentrations of 0.1–0.7 ng/L. In comparison to the water concentrations, the levels of Hg in fish tissue indicate bioaccumulation in the food chain to the trophic levels of forage fish and piscivorous fish. Although elevated Hg concentrations were observed in feathers, poor correlations among feathers, blood, tissues, and eggs were found. Therefore, Caldwell, Arnold, and Gould (1999) suggested that Hg concentrations in feathers alone would not be useful in predicting risks to double-crested cormorants.

Table 4-13. Mercury Concentrations (µg/g ww) in Double-Crested Cormorant Nestlings from Two Reservoirs on the Rio Grande River.

Tissue	Elephant Butte[a]	Caballo
Eggs	0.23 ± 0.01 (36)	0.30 ± 0.01 (36)
1st blood collection	0.36 ± 0.03 (17)	0.36 ± 0.03 (18)
2nd blood collection	0.34 ± 0.02 (10)	0.39 ± 0.01 (3)
Liver	0.40 ± 0.02 (12)	0.40 ± 0.02 (7)
Muscle	0.17 ± 0.11 (12)	0.18 ± 0.01 (7)
Primary feathers	2.42 ± 0.32 (12)	3.54 ± 0.35 (7)
Secondary feathers	2.06 ± 0.21 (12)	4.89 ± 0.97 (7)
Tail feathers	2.34 ± 0.25 (12)	4.01 ± 0.67 (7)

Source: Caldwell, Arnold, and Gould 1999.
[a]Average ± standard error (sample number).

Wolfe and Norman (1998) conducted a field validation study to test the rigor of the Great Lakes Water Quality Criteria Initiative (GLWQI) predictive wildlife criteria. They sampled the sediment, water, animal food, and piscivorous birds and mammals at Clear Lake, California. Clear Lake was chosen because it is contaminated with Hg and MeHg and because the wildlife species are different from those used to derive the GLWQI predictive model. Clear Lake is California's largest natural lake and is located in an area with cinnabar-bearing serpentine soils. Clear Lake has been contaminated with Hg from open-pit cinnabar mining at the Sulphur Bank Mercury Mine (SBMM) on the Oaks Arm of Clear Lake. Near the former mine, total sediment Hg was 40–370 mg/kg with an average of 220 mg/kg. In unfiltered deep water in Oaks Arm, total Hg was 81 ng/L and 22 ng/L in surficial water. Concentrations of MeHg in Oaks Arm sediment were 10 ng/g and 0.1–1.0 ng/L in unfiltered deep water. Great blue heron, double-crested cormorant, red-winged blackbird (*Agelaius phoeniceus*), Brewer's blackbird (*Euphagus cyanocephalus*), and cliff swallow (*Hirundo pyrrhonota*) nestlings were collected. Raccoons (*Procyon lotor*) and mink (*Mustela vison*) were trapped. Mercury concentrations were determined in blood, brain, muscle, kidney, liver, claws, and feathers or fur. The primary fish fed to heron and cormorant nestlings was the Sacramento hitch (*Lavinia exilicauda*), a trophic level 2 (TL-2) species that feeds on *Daphnia* and other zooplankton. The Hg concentrations in whole Sacramento hitch were approximately 0.3–0.85 µg/g, which gave a dietary intake of 0.073–0.21 mg/kg-day for great blue heron nestlings. Mean Hg concentrations in great blue heron nestling tissues were 0.35 µg/g ww in brain, 1.46 µg/g ww in liver, 2.36 µg/g ww in feathers, 1.18 µg/g ww in blood, and 1.09 µg/g ww in kidney. The correlations between blood Hg and brain, kidney, and liver Hg were significant. In the first year of sampling, no correlation was found between feather Hg and either liver or brain Hg. In the second year of sampling, using improved feather sampling and analysis techniques, a significant correlation was found between feather and liver Hg, but not between feather and brain Hg. In contrast, a strong correlation between feather Hg and both liver and brain Hg was observed in cormorants. Concentrations of Hg in the three insectivorous passerine bird species were 0.018–0.03 µg/g ww in brain, 0.094–0.364 µg/g ww in feathers, and 0.24–0.92 µg/g ww in liver. Insects collected in sweep samples had Hg concentrations of 0.01–0.031 µg/g ww, and insects from dip samples had Hg concentrations of 0.254–0.420 µg/g ww. Raccoon (*Procyon lotor*) Hg concentrations were variable. The highest Hg concentrations measured, 1.15 µg/g ww in brain and 8.46 µg/g ww in liver, were below levels of concern for Hg toxicity in wild mammals. Only eight mink were trapped in two years of sampling. Concentration of Hg in mink tissues (μg/g ww) ranged from 0.12 to 7.1 in brain, 0.44 to 10.1 in liver, 2.13 to 21.5 in fur, 0.08 to 0.63 in blood, 0.28 to 1.13 in muscle, 0.41 to 1.3 in kidney, and 2.28 to 34.1 in claws. After omitting data from one mink with an exceptionally high brain Hg concentration, a significant correlation between fur Hg and brain Hg was observed, but fur Hg was not correlated with blood or liver Hg. Although fur or feathers can be sampled easily in mammals or birds, the relationship to Hg concentrations in other tissues is not well defined. Based on the total Hg concentration in unfiltered water in Oaks Arm of Clear Lake, Wolfe and Norman (1998) calculated a BAF of 11,000 L/kg for TL-2 based on Sacramento hitch, and a BAF of 31,200 L/kg for TL-3 based on great blue heron liver. Wolfe and Norman (1998) also estimated a mixed TL-4 BAF of 190,600 L/kg based on mink. These values were similar to the BAFs of 27,900 L/kg for TL-2 and 140,000 L/kg for TL-4 used in developing the GLWQI wildlife value of 1300 pg/L.

CHAPTER 5.0

TOXICITY

Methylmercury is highly toxic due to its affinity for sulfur-containing organic compounds (proteins). Methylmercury forms covalent bonds with sulfhydryl groups of proteins and other macromolecules (U.S. EPA 1997a), which allows it to migrate readily through biological membranes (U.S. EPA 1997a). Thus, methylmercury can easily reach sensitive tissues such as the brain or migrate across the placenta to reach the fetus.

The toxicity of mercury is controlled by dose, form of mercury, various physical and chemical parameters (pH, temperature, DOC), exposure to other stressors, life stage, age, sex, species, physical condition, and the ability to detoxify mercury (U.S. EPA 1997a).

Several target tissues are affected by mercury, but the central nervous system (CNS) is the primary target in fish, birds, and mammals (U.S. EPA 1997b; Wiener and Spry 1996). In adult humans, mercury is thought to affect astrocytes and other neuroglial cells. In human embryos, adverse effects are due to the disturbances in antimitotic activity.

5.1 Toxicokinetics and Toxicodynamics

5.1.1 Fish

Methylmercury rapidly diffuses across the gills (7% to 12% of MeHg [Wiener and Spry 1996]) and enters the blood stream. Methylmercury is also rapidly accumulated through the gut (65% to 80%) and transferred to the blood stream where it is subsequently transported to all organs including the liver, kidney, and brain (Wiener and Spry 1996). Methylmercury is ultimately deposited in the muscle tissue (Wiener and Spry 1996).

5.1.2 Terrestrial Biota

Methylmercury from diet is efficiently (65% to 100% [Wiener and Spry 1996]) absorbed from the gut and rapidly associated with sulfhydryl groups in blood (free amino acids such as cysteine and glutathione). These mobile complexes transport mercury to tissues and organs and may facilitate movement across cell membranes. There is evidence that this complex can also transport mercury across the brain and placental barriers (U.S. EPA 1997a).

Inorganic mercury is absorbed through the gut and fish gills but at a much lower efficiency and is excreted much more readily (Wiener and Spry 1996). Hence, the levels of inorganic mercury in tissues tend to be lower than that of organic mercury.

5.2 Symptoms

5.2.1 Fish

Effects of mercury poisoning in fish include death, reduced reproduction, impaired growth and development, behavioral abnormalities, altered blood chemistry, impaired osmoregulation, reduced feeding rates and predatory success, and altered respiration (U.S. EPA 1997a).

Methylmercury can also act as an endocrine disruptor and has the potential to adversely affect reproductive success in fish (Fynn-Aikins *et al.* 1998).

5.2.2 Terrestrial Biota

Adverse effects of mercury exposure on terrestrial species include death, decreased reproduction, decreased growth, decreased immune response, inhibition enzyme induction, change in behavior, and change in respiration (U.S. EPA 1997a).

In humans, the most sensitive indicator of mercury poisoning is cognitive impairment in individuals exposed during development (U.S. EPA 1997a). However, this indicator cannot be easily adapted to laboratory or wildlife species (U.S. EPA 1997a). Instead, reduced hatching success, reproduction and mobility are used as endpoints.

Acute effects include muscular incoordination, falling, slowness, fluffy feathers, calmness, withdrawal, hyperactivity, and eyelid droop (U.S. EPA 1997a). Chronic effects include liver damage, kidney damage, neurobehavioral effects, reduced food consumption, weight loss, spinal cord damage, enzyme system damage, reduced cardiovascular function, impaired immune response, reduced muscular coordination, impaired growth and development, altered blood and serum chemistry, and reproductive effects. The reproductive and behavioral effects are of primary concern because they occur at dietary levels below those associated with overt toxicity (U.S. EPA 1997a).

The most common assessment endpoint for individuals and populations include mortality, growth, development, and reproduction (U.S. EPA 1997a).

5.3 Toxicity to Aquatic Life

5.3.1 Dietary Exposure

Concentrations of methylmercury in freshwater, even at contaminated sites, are not high enough to cause mercury toxicity (Wiener and Spry 1996). Thus, the major concern is the dietary uptake and subsequent accumulation of methylmercury in tissues. For this reason, this report focuses on reviewing the results from studies that reported levels of methylmercury in diet and in tissues associated with adverse effects. Unfortunately, such data are limited in the literature. The existing data tend to favor fish. Moreover, the information is spread out over several decades invoking concerns about data quality due to changes in testing and analytical protocols since the 1970s.

5.3.1.1 Walleye (*Stizostedion vitreum*)

Scherer, Armstrong, and Nowak (1975) considered the effects of dietary exposure to methylmercury on juvenile walleye fed contaminated northern pike (*Esox lucius*) for 314 days. The fish in the treatment group were fed a diet containing 7.9 mg of MeHg/kg ww. The food (northern pike) was collected from a contaminated site in Clay Lake, Ontario. The fish in the control group were fed a diet containing 0.41 mg MeHg/kg ww. Because the daily food intake rate was about 2 g per fish, the daily dose through feed was about 0.016 mg MeHg/day in the treatment group and 0.0008 mg MeHg/day in the control group.

The fish exposed to the contaminated diet became emaciated, had lower locomotor activity, and showed reduced coordination, escape behavior, and response to light. The mortality rate was increased compared to the controls.

Scherer, Armstrong, and Nowak (1975) also reported tissue concentrations of methylmercury in fish that died during the bioassay. Twenty fish died between day 42 and day 63 of the test. In comparison, only two fish died in the controls. Therefore, the observed mortality was probably due to methylmercury poisoning. The concentrations of total mercury in the dead walleye ranged from 3 to 6 mg MeHg/kg ww in the brain tissue, 6 to 14 mg MeHg/kg ww in the liver, and 5 to 8 mg MeHg/kg ww in the muscle.

The total mortality at the end of the 314-day test was 88%. The cumulative mortality in the controls was 27%. The tissue residues of mercury in walleye that died after day 240 ranged from 15 to 40 mg MeHg/kg ww in the brain tissue, 18 to 50 mg MeHg/kg ww in the liver, and 15 to 45 mg MeHg/kg ww in the muscle. The tissue residues in the control fish were less than 2 mg MeHg/kg ww in the brain tissue, 3 mg MeHg/kg ww in the liver, and 2.5 mg MeHg/kg ww in the muscle.

5.3.1.2 Rainbow Trout (*Oncorhynchus mykiss*)

Wobeser (1975) examined the effects of dietary exposure to methylmercury on fingerling rainbow trout. Fish weighing 17 to 52 g were fed a commercial feed spiked with methylmercuric chloride for 105 days. The nominal treatment concentrations were 4, 8, 16, and 24 mg MeHg/kg ww. The last two treatments reduced the mean growth rate in fish. The muscle burdens associated with this effect ranged between 12 and 23 mg MeHg/kg ww. The muscle tissue levels in the control fish were less than 0.2 mg MeHg/kg ww. No differences in mortality, appetite, capture avoidance, or visual acuity was noted for the other treatments.

Dietary exposure to rainbow trout was also studied by Rodgers and Beamish (1982). The trout were fed a commercial trout food spiked with methylmercuric chloride for 84 days. The nominal concentrations in the feed were 0, 25, 45, 75, and 95 mg MeHg/kg ww. The doses of 75 and 95 mg MeHg/kg ww caused darkened skin and lethargy in 20% to 30% of the fish. Levels of total mercury in whole fish associated with these doses and effects were between 30 and 35 mg MeHg/kg ww. Doses of 25, 45, and 95 mg MeHg/kg ww reduced the appetite of fish. Levels of total mercury in whole fish at these doses were 10 to 30 kg MeHg/kg ww during days 56 to 84 of the test.

The growth rates of rainbow trout were inversely correlated to the methylmercury concentration in the diet. The depressed growth was due to the reduced appetite and to the lower digestibility of the contaminated diet. Other endpoints such as serum osmolality, sodium, and potassium levels; histology of the gill, liver, kidney, and intestine; and mortality were not altered by exposure to methylmercury.

Matida *et al.* (1971) fed mercury-contaminated shellfish collected from Minamata Bay, Japan, to rainbow trout for 270 days. The shellfish contained 48 mg Hg/kg dw. About 30% of the total mercury was in the methylated form. The fish showed loss of appetite, reduced activity, darkened skin, and loss of equilibrium. The exposed fish also had reduced visual acuity, were more likely to collide with the tank walls, and had increased difficulty catching food. The trout grew more slowly than the controls. The tissue levels were 19 to 48 mg/kg ww in the whole body, 20 to 28 mg/kg ww in the muscle, 26 to 68 mg/kg ww in the liver, and 16 to 30 mg/kg ww in the brain.

5.3.2 Waterborne Exposures

5.3.2.1 Brook Trout (*Salvelinus fontinalis*)

McKim *et al.* (1976) examined the toxicity of waterborne methylmercury (methylmercuric chloride) to chronically exposed brook trout. Fish exposed to 0.00293 mg MeHg/L showed loss of appetite, muscle spasms, and deformities. Many of the fish died (88%) soon after. The tissue concentrations of methylmercury in the dead fish averaged 155 mg/kg ww in the spleen, 58 mg/kg ww in the liver, 147 mg/kg ww in the kidney, 48 mg/kg ww in the gill, 42 mg/kg ww in the brain, 32 mg/kg ww in the gonad, and 24 mg/kg ww in the axial muscle and the whole body. The variability in brain concentrations was relatively low and the brain concentration was independent of duration of exposure.

Fish exposed to 0.00093 μg Hg/L for 39 weeks had sluggish behavior, deformities, reduced growth, and mortality. The tissue concentrations associated with the given exposure and observed effects averaged 39 mg/kg ww in the spleen, 24 mg/kg ww in the liver, 27 mg/kg ww in the kidney, 22 mg/kg ww in the gill, 17 mg/kg ww in the brain, 12 mg/kg ww in the gonad, 10 mg/kg ww in the axial muscle, and 9 mg/kg ww in the whole body.

Concentrations of 5 to 7 mg/kg ww in the whole body were associated with toxic symptoms and death. However, no deaths or abnormalities were observed at 2.7 mg/kg ww. The no-observed-effect concentration for methylmercury in water was reported to be 0.29 μg Hg/L. The tissue levels of methylmercury at this exposure averaged 12 mg/kg ww in the spleen, 8 mg/kg ww in the liver, 9 mg/kg ww in the kidney, 6 mg/kg ww in the gill, 5 mg/kg ww in the brain, 3 mg/kg ww in the gonad, and 5 mg/kg ww in the muscle.

5.3.2.2 Rainbow Trout

Niimi and Kissoon (1994) measured residues of methylmercury in subadult rainbow trout killed by exposure to waterborne methylmercuric chloride. The concentrations of mercury in the brain tissue of rainbow trout that died during the test varied 50-fold among individuals (from 0.6 to 32 mg/kg ww). In contrast, mercury levels in the gills varied little and averaged 66, 51, 64, and 56 mg/kg ww in fish exposed to 4, 10, 13, and 34 μg Hg/L. The last three treatments were sufficiently high to cause severe adverse effects on the gills. The lower exposure concentrations caused the test fish to become lethargic and feed less. Whole-body concentrations in fish exposed to 0.009 mg Hg/L until death ranged from 4 to 27 mg/kg ww (average of 11 mg/kg ww).

Niimi and Kissoon (1994) observed that mercury concentrations in the kidney, liver, spleen, brain, and the muscle of dead rainbow trout increased with increasing exposure time. This suggested that death did not occur at a single threshold level. Thus, the critical tissue concentration concept may not apply to acute lethal exposures. They suggested that the critical concentration approach might be more appropriate if based on whole-body residues. In earlier work, McKim *et al.* (1976) found that mercury concentrations in the brains of brook trout killed by exposure to 2.9 µg Hg/L of waterborne methylmercury varied little and was independent of exposure duration.

Niimi and Lowe-Jinde (1984) showed that subadult rainbow trout exposed to waterborne methylmercury at 0.15 µg Hg/L (as methylmercuric chloride) for 75 days accumulated whole-body concentrations of 12 mg/kg without apparent effects on survival or hematological parameters.

5.3.2.3 Fathead Minnows (*Pimephales promelas*)

Olson *et al.* (1975) exposed fathead minnows to waterborne methylmercuric chloride for 336 days. The treatment concentrations ranged from 0.018 to 0.247 µg Hg/L. The mean concentrations of mercury in whole fathead minnows after the test ranged from 0.21 mg/kg ww in controls to 11 mg/kg in fish exposed to 0.247 µg Hg/L treatment. There were no readily apparent effects on the survival, behavior, growth, or general appearance of the fish.

5.3.3 Field Observations

5.3.3.1 Minamata Fish

Kitamura (1968) collected and analyzed six "weakened" fish species from Minamata Bay, Japan. These fish were lethargic and could be easily captured by hand suggesting depressed locomotor activity and escape behavior. The fish were frequently emaciated and had lesions in the brain. Mercury con-centrations in the muscle of these fish ranged from 8.4 to 24 mg/kg (average 15 mg/kg ww).

5.3.3.2 Northern Pike (*Esox lucius*)

Lockhart *et al.* (1972) examined the physiological and biochemical condition of mercury-contaminated northern pike from a polluted site in Clay Lake, Ontario. The fish were emaciated, had low hepatic fat stores, had low levels of total protein, glucose, and alkaline phosphatase in blood serum, and had low serum cortisol levels. The concentrations of methylmercury ranged from 6 to 15 mg/kg ww in the muscle. The fish were between three and eight years old.

5.3.3.3 Rock Bass (*Ambloplites rupestris*)

Bidwell and Heath (1993) studied the health of rock bass collected from contaminated sites on the South River, Virginia, relative to the residues of mercury in axial muscle and liver tissues. The residues of total mercury (most likely methylmercury) in contaminated fish averaged 1.4 mg/kg ww in muscle and 2.9 mg/kg ww in liver. These residues, however, did not affect any physiological or biochemical variables that would affect fish health.

5.3.4 Mercury Effects on Early Life Stages

Although the literature contains many manuscripts on the effects of waterborne inorganic and organic mercury on early life stages, they will not be reviewed in detail in this report, because waterborne concentrations of inorganic mercury and methylmercury known to cause lethal or sublethal (teratogenic) effects in young fish, embryos, eggs or sperm are typically orders of magnitude greater than the recently measured concentrations in freshwaters.

5.3.4.1 Walleye

Whitney (1991) studied the effects of maternal transfer of mercury on the hatching success, survival, growth, and behavior of embryo and larval walleye from two lakes in Wisconsin. Eggs were taken from six fish from Escanaba Lake (range in total length, 38 to 54 cm; range in age, five to seven years) and from six fish from nearby Trout Lake (48 to 59 cm; five to seven years). The lakes in the area contained between 2.1 and 7 ng Hg/L and 0.05 to 0.33 ng MeHg/L. Therefore, the fish were exposed to very low levels of mercury.

The residues of mercury in the collected eggs ranged from 0.002 to 0.013 mg/kg ww (Escanaba Lake) and from 0.015 to 0.058 mg/kg ww (Trout Lake). Mercury concentrations in the eggs from the Trout Lake were positively correlated with those in the maternal muscle tissue. The residues in the muscle ranged from 0.20 to 0.57 mg/kg ww. The residues in the eggs from the Escanaba Lake were negatively correlated with those in maternal muscle tissue. The levels in the maternal muscle tissue ranged from 0.14 to 0.36 mg/kg ww.

The hatching success of the collected eggs ranged from 81% to 100% and was negatively correlated with the egg residue of mercury. The survival of embryos ranged from 70% to 99% and was also negatively correlated with mercury residues in eggs. The growth, survival, and the behavior of larval fish were not correlated with mercury residue levels in eggs.

5.3.4.2 Rainbow Trout

Birge *et al.* (1979) studied mercury residues in fertilized eggs of rainbow trout associated with embryo mortality. Residues of 0.068 mg/kg ww after four days of exposure and 0.097 mg/kg ww after 7.5 days of exposure were associated with survival of 79% to 83% and 23% to 26%, respectively. The mortality in the embryos occurred at mercury concentrations of about 0.07 to 0.10 mg/kg ww in the fertilized eggs.

5.3.4.3 Brook Trout

McKim *et al.* (1976) exposed three generations of brook trout to waterborne methyl-mercuric chloride. The eggs from the females contained 12.5 mg/kg ww. The hatched offspring were deformed and died. The third generation of embryos was also deformed. The egg residue level associated with this effect was 2.2 mg/kg ww.

5.3.5 Endocrine Disruption Activity

5.3.5.1 Largemouth Bass (*Micropterus salmoides*)

Fynn-Aikins *et al.* (1998) exposed adult male and female largemouth bass to methylmercury chloride in diet for two weeks. Plasma estradiol or 11-ketotestosterone levels were not affected at the dose of 1.56 mg MeHg/kg of diet. However, the dose of 3.12 mg MeHg/kg diet did reduce plasma hormone levels in males and females. The authors concluded that the dietary exposure to methylmercury depresses hormone levels. They also indicated that methylmercury acts as an endocrine disrupter and might impair reproduction in fish.

5.4 Toxicity to Terrestrial Biota

The literature contains many laboratory-based studies on the toxicity of mercury to various animals. However, most of these data are difficult to extrapolate to the field because the animals were challenged with acute doses of methylmercury that are unlikely to be found in the field. Thus, caution is required when using laboratory results for the purpose of estimating mercury toxicity and deriving exposure limits. The results from existing field and laboratory studies are reviewed below.

5.4.1 Avian Toxicity

5.4.1.1 Mallards (*Anas platyrhynchos*)

Heinz (1974) reported a sublethal effect of mercury on mallard ducks (*Anas platy-rhynchos*). Adult birds were fed a diet containing 3 mg/kg of methylmercury (dw) (methyl-mercury dicyandiamide) for one year. The exposed individuals appeared healthy. However, Heinz (1974) observed reduced egg output, reduced hatching success, as well as increased mortality of ducklings. The dead ducklings had brain lesions (Heinz and Locke 1976). The levels of mercury in eggs were more than 5 mg/kg ww.

In the second breeding year, egg production and hatching success were not adversely affected. The egg residues dropped to 1.0 mg/kg ww (Heinz 1976).

5.4.1.2 While-Tailed Eagles (*Haliaeetus albicilla*)

Henriksson *et al.* (1966) reported mercury concentrations of 4.6 to 27.1 in the liver and 48.6 to 123.1 mg/kg ww in the kidney of white-tailed eagles found dead. The death of the birds was attributed to mercury poisoning. Koeman, Hadderingh, and Bijlveld (1972) reported a mercury concentration of 48.2 mg/kg ww in the liver of a white-tailed eagle also found dead. Oehme (1981) stated that mercury poisoning was the cause of death for several white-tailed eagles found dead in Germany. The median mercury concentration in the liver was 91 mg/kg ww. The kidney contained 120 mg/kg ww. Falandysz (1986) and Falandysz, Jakuczun, and Mizera (1988) recorded mercury concentrations ranging from 11 to 33 mg/kg ww in the liver of three white-tailed eagles that likely died from mercury poisoning.

Koivusaari *et al.* (1980) reported that mercury concentrations ranging from 1.31 to 4.16 mg/kg dw (0.33 to 2.25 mg/kg ww) in eggs of white-tailed eagles had no effect on reproductive success. Helander, Olsson, and Reutergardh (1982) found no correlation between mercury concentration in eggs of white-tailed eagles from Sweden and reproductive success. The eggs contained on average 4.6 mg/kg dw.

5.4.1.3 Bald Eagles (*Haliaeetus leucocephalus*)

Wiemeyer *et al.* (1984) observed an inverse relationship between the concentration of mercury in bald eagle eggs and mean five-year productivity. However, the relationship was weak. DDE was the most important variable driving the relationship. Mercury concentrations in the unhatched eggs were seldom more than 1.0 mg/kg ww. The geometric mean concentration was 0.4 mg/kg ww (Wiemeyer *et al.* 1984).

5.4.1.4 Golden Eagles (*Aquila chrysaetos*)

Furness *et al.* (1989) observed that decreased reproductive success was frequently associated with increased mercury concentration in the feathers of golden eagles. These eagles came from western Scotland and the mean feather residues were about 8 mg/kg fw.

Newton and Galbraith (1991) observed a similar trend between the decreased reproductive success and mercury residue levels in golden eagle eggs. However, authors from both studies were uncomfortable linking mercury residues with reproductive success, because other natural parameters such as diet could have played a major role in affecting reproduction.

5.4.1.5 Kestrel (*Falco sparverius*)

Koeman, ten Noever de Brauw, and de Vos (1969) measured mercury concentrations ranging from 73 to 125 mg/kg ww in the kidneys of kestrels. At these levels, the birds were dead or showed signs of mercury poisoning. Koeman *et al.* (1971) fed kestrels with mice spiked with methylmercury dicyandiamide (whole body concentration of 13.3 mg/kg ww). Mercury residue levels in the liver of the dead birds ranged from 49 to 122 mg/kg ww. Peakall and Lincer (1972) did not observe any eggshell thinning in eggs laid by kestrels dosed with 10 mg/kg of dimethylmercury or methylmercury chloride orally or by injection.

5.4.1.6 Loon (*Gavia immer*)

Fimreite (1974) found that mercury contamination of Ball Lake, Ontario was linked to the absence of loon chicks at this site. Mercury concentrations in the liver were as high as 90.5 mg/kg ww. The mean concentration was 51.9 mg/kg. Barr (1986) reported an inverse relationship between reproductive success in common loons and mercury concentration. He indicated that concentrations of 0.3 to 0.4 mg/kg ww in the prey were sufficient to impair territorial fidelity and egg laying in loons.

Elevated prey fish Hg concentrations of 0.3–0.4 μg/g ww have been associated with reduced reproductive success in loons. Scheuhammer *et al.* (1998) reported that loons were absent from several of the lakes with fish Hg concentrations in excess of 0.4 μg/g.

5.4.1.7 Common Terns (*Sterna hirundo*)

Fimreite (1974) found depressed hatching success in common tern eggs containing 3.65 mg/kg ww. The eggs were collected from Ball Lake, Ontario. Hatching success at another location (Wabigoon Lake) was high. The eggs at this location contained on average 1.0 mg/kg. Connors *et al.* (1975) concluded that the low reproductive success of terns living in Hamilton Harbor, Ontario was not related to mercury contamination. The average liver residues were 2.28 mg/kg ww and the maximum residue was 4.74 mg/kg ww.

5.4.1.8 Pheasants (*Phasianus colchicus*)

Gardiner (1972) found that a feeding regime of 35 days and a diet containing 33 mg/kg methylmercury resulted in 90% mortality in pheasants. Spann *et al.* (1972) reported that a diet containing 12.5 mg/kg ww ethylmercury p-toluene sulfonanilide was lethal to pheasants. However, a diet containing 4.2 mg/kg did not have any lethal effects in adults. The lower dose did, however, depress egg production and survival of third-week embryos. The mean mercury concentration in eggs was 1.5 mg/kg ww. Borg *et al.* (1969) discovered that mercury concentrations of 1.3 to 2.0 mg/kg ww in pheasant eggs reduced hatching success. Fimreite (1971) reported that adult birds fed a diet containing 2 to 3 mg/kg of methylmercury dicyandiamide for 12 weeks did not show any signs of poisoning. There were no mortalities, depressed feeding rates, altered mating behavior or effects on egg production. However, this dose resulted in the production of eggs without shells. This dose also led to a lower egg weight, decreased hatching success, and a lower fertilization rate. Mercury concentrations in the unhatched eggs ranged from 0.5 to 1.5 mg/kg ww. The liver of adult birds contained 2 mg/kg mercury ww.

5.4.1.9 Goshawks (*Accipitridae*)

Borg *et al.* (1970) fed juvenile goshawks a diet composed of chicks that were spiked withmethylmercury dicyandiamide. The levels of methylmercury in these chicks were 10 and 40 mg/kg ww in the muscle and liver, respectively. The average mercury intake delivered to the goshawks was 20 mg. The exposure caused the birds to feed less, lose weight, and die within 30 to 47 days. The dead birds were emaciated and showed substantial reductions in the fat and muscle volumes. Methylmercury residues in the birds ranged from 103 to 144 mg/kg ww in the liver and 121 to 138 mg/kg in the kidney.

5.4.1.10 Chickens (*Gallus domesticus*)

Scott *et al.* (1975) exposed chickens to inorganic mercury (mercuric sulfate or mercuric chloride) in the diet. Some doses were as high as 200 mg/kg. However, they observed little or no effects on egg weight, fertility, and hatching success. They also exposed another set of birds to organic mercury as methylmercuric chloride in diet. The dietary intake ranged from 10 and 20 mg/kg. This time, there were severe adverse effects observed in terms of reduced egg weight, depressed fertility, lower hatching success. The birds also exhibited severe ataxia and produced misshapen and truncated eggs that featured thin shells.

Parkhurst and Thaxton (1973) exposed young chickens to mercuric chloride in drinking water at doses reaching 250 µg/mL. The exposed birds responded with depressed growth and increased mortality. Gardiner (1972) found that a diet containing 33 mg/kg of methylmercury was lethal to chickens after 35 days of exposure.

5.4.1.11 Buzzards (*Accipitridae*)

Koeman, ten Noever de Brauw, and de Vos (1969) measured mercury concentrations ranging from 85 and 93 mg/kg ww in the kidneys of dead or dying buzzards.

5.4.1.12 Long-Eared Owl (*Asio otus*)

Koeman, ten Noever de Brauw and de Vos (1969) measured mercury concentrations ranging from 68 mg/kg ww in the kidneys of dead or dying owl.

5.4.1.13 Turkey Vulture (*Cathartes aura*)

Fimreite (1974) indicated that concentrations of mercury in vulture livers, ranging from 40 to 121 mg/kg ww, were associated with impaired wing movement.

5.4.1.14 Herring Gulls (*Larus argentatus*)

Gilman *et al.* (1977) did not find any adverse effects of egg residues ranging from 0.22 to 0.51 mg/kg ww. The herring gull eggs were collected from the Great Lakes. Vermeer, Armstrong, and Hatch (1973) reported that the hatching success of herring gull eggs was not affected by mercury concentrations in eggs ranging from 2.3 to 15.8 mg/kg ww.

5.4.1.15 Forster's Tern (*Sterna forsteri*)

King, Custer and Quinn (1991) reported that there was no correlation between the mercury concentration in tern eggs and the hatching success at a contaminated site in Texas. The mean residue in eggs was 0.40 mg/kg ww.

5.4.1.16 Black Skimmers (*Rynchops niger*)

King, Custer and Quinn (1991) did not find any correlation between the mercury concentration in eggs of black skimmers and the hatching success at the site in Texas. The average residue in eggs was 0.46 mg/kg ww.

5.4.1.17 Prairie Falcon (*Falco mexicanus*)

Fyfe, Risebrough and Walker (1976) investigated the effects of mercury residues on eggs and reproductive success. Mean egg residues of 0.35 mg/kg dw had no effect on reproduction.

5.4.1.18 Richardson's Merlin (*Falco columbarius*)

Fyfe, Risebrough, and Walker (1976) also found no significant relationships between the mercury concentration in eggs of merlin and reproductive success. The average egg residue level was 0.66 mg/kg dw. However, Newton and Haas (1988) found that mercury concentrations above 3 mg/kg dw reduced the brood size of merlin eggs.

5.4.1.19 Marsh Harriers (*Circus aeruginosus*)

Odsjo and Sondell (1977) reported that mercury concentrations in eggs ranging from 1 to 2 mg/kg dw did not affect the number of young in a clutch.

5.4.1.20 Peregrine Falcon (*Falco peregrinus*)

Newton, Bogan, and Haas (1989) reported that brood size was affected by egg residue levels ranging from 0.21 mg/kg dw for inland sites to 1.27 mg/kg dw for coastal sites.

5.4.1.21 White-Faced Ibis (*Plegadis chihi*)

Henny and Herron (1989) found that mercury concentrations in eggs below 1.0 mg/kg dw had no effect on bird reproduction.

5.4.1.22 Great Skuas (*Catharacta skua*)

Thompson, Hamer, and Furness (1991) found no effects of mercury residues in the liver on a range of reproductive parameters and survival of skuas. The mercury concentrations in the liver ranged from 4.5 to 33.4 mg/kg dw. Methylmercury concentrations in the liver ranged from 1.8 to 11.1 mg/kg dw.

5.4.1.23 Grackles (*Quiscalus quiscula*)

Finley, Stickel, and Christensen (1979) fed grackles a diet with 40 mg/kg of methylmercury dicyandiamide. This dietary concentration was lethal and the birds died within six days. The liver of the dead birds contained 54.5 mg/kg ww.

5.4.1.24 Starlings (*Sturnus vulgaris*)

Finley, Stickel, and Christensen (1979) also fed starlings a diet with 40 mg/kg of methylmercury dicyandiamide. This dietary concentration killed starlings within eight days.

Nicholson and Osborn (1984) reported that 1.1 mg/kg of mercury dw in starling diet resulted in kidney lesions. The kidney residue was 36.3 mg/kg dw. The liver residue was 6.55 mg/kg dw.

5.4.1.25 Cowbirds (*Molothrus*)

Finley, Stickel, and Christensen (1979) exposed cowbirds to 40 mg/kg methylmercury dicyandiamide in diet. This dietary concentration was lethal within 10 days.

5.4.1.26 Red-Winged Blackbirds (*Agelaius phoeniceus*)

Finley, Stickel and Christensen (1979) exposed blackbirds to a diet containing 40 mg/kg methylmercury dicyandiamide. The dietary concentration was lethal, and the birds died within 11 days. The liver residue in dead birds was 126.5 mg/kg.

5.4.1.27 Red-Tailed Hawks (*Buteo jamaicensis*)

Fimreite and Karstad (1971) exposed one-year-old red-tailed hawks to a diet of chicks spiked with methylmercury dicyandiamide. The methylmercury residues in the chick livers ranged from 3.9 to 10.0 mg/kg ww. The four-week exposure to a concentration of 10 mg/kg ww in diet resulted in nerve lesions, slight weight loss, loss of coordination, and death. Methylmercury concentrations in the livers of hawks ranged from 16.7 to 20.0 mg/kg ww.

5.4.1.28 Black Ducks (*Anas rubripes*)

Finley and Stendell (1978) exposed adult black ducks to a diet containing 3 mg/kg dw of methylmercury (methylmercury dicyandiamide). The treatment lasted two years. The treated birds showed no overt adverse effects. However, clutch size and the egg hatching success were lower than those found in the controls. Duckling survival was reduced. Nerve lesions were also observed. The mean methylmercury concentrations in the eggs ranged from 4.70 to 5.53 mg/kg ww.

5.4.1.29 Japanese Quail (*Coturnix japonica*)

Thaxton and Parkhurst (1973) supplied Japanese quail with drinking water spiked with 125 mg/kg of mercuric chloride. The affected birds showed reduced mating attempts, egg fertility, egg hatching success, ova development, and testes weight.

Gullvag, Aagdal, and Eskeland (1978) recorded damage to liver littoral cells in quail fed 8 mg/kg ww of methylmercury chloride for six weeks. Stoewsand *et al.* (1971) observed thinning of quail eggshells after the birds were fed a diet containing 8 mg/kg of mercuric chloride.

5.4.1.30 Ring Doves (*Streptopelia risoria*)

Peakall and Lincer (1972) found that a dietary concentration of 10 mg/kg of dimethylmercury or methylmercury chloride given orally or via injection had no effects on the thinning of eggshells of doves.

5.4.2 Summary of Effect Levels in Birds

The analysis of published data on mercury toxicity to birds revealed that mercury concentrations in the liver and the kidney higher than 30 mg/kg ww are lethal to several species. Mercury concentrations in eggs above 0.5 mg/kg ww (2.5 mg/kg dw at egg water content of 80%) begin to have detrimental effect on reproduction. Dietary mercury concentrations of 10 mg/kg ww are lethal to birds of prey, passerines, and pheasants. Dietary concentrations above 1 mg/kg ww can have detrimental effects on reproduction.

Marine seabirds and piscivorous inland species that may be exposed to higher mercury levels are generally more tolerant of higher exposures (EPA 1997a), perhaps because marine and freshwater ecosystems have naturally richer mercury burdens and these birds have various adaptations for dealing with excess mercury (Honda *et al.* 1990). For example, marine birds can effectively transform the toxic methylmercury into the less toxic inorganic mercury, an adaptation that many inland birds do not have (Honda *et al.* 1990).

5.4.3 Mammalian Toxicity

5.4.3.1 Otter (*Lutra canadensis*)

Wren (1985) analyzed the mercury content in the liver and the kidney of an otter found dead at Clay Lake, Ontario. The animal was thought to have succumbed to mercury poisoning. The total mercury residues were 96 mg/ kg ww in the liver and 58 mg/kg ww in the kidney. The levels in the muscle tissues and the brain were 36 and 30 mg/kg, respectively. Kruuk and Conroy (1991) indicated that high residues of mercury in otter livers (65 mg/kg dw) may have a detrimental effect on their survival. O'Connor and Nielsen (1980) fed otters diets containing methylmercury at 2, 4, or 8 mg/kg ww as methylmercury hydroxide. The feeding regime was equivalent to 9.3, 17.0, and 37.0 mg/kg/day and the mean survival times were 184, 117, and 54 days. The mean mercury concentrations in the livers of dead otters were 33.0 mg/kg ww. The kidney concentration was 39.0 mg/kg ww, and the brain concentration was 19.0 mg/kg ww. Dead otters also had lesions in the central nervous system.

5.4.3.1 Mink (*Mustela vison*)

Wobeser (1976) collected and analyzed a mink that was behaving erratically near the South Saskatchewan River. The animal died soon after it was collected and was subjected to a post mortem examination. The carcass contained no visible lesions. However, the cerebral cortex and the cerebellum contained many lesions. The liver contained 58.2 mg/kg ww of mercury and the kidney contained 31.9 mg/kg ww of mercury. The levels in the muscle tissue, brain, and fur were 15.2, 13.4, and 34.9 mg/kg, respectively.

Aulerich, Ringer, and Iwamoto (1974) fed adult mink a diet containing 5 mg/kg ww of methylmercury. The mink developed signs of mercury poisoning after 24 days of exposure and died 30 days after the beginning of the experiment. The methylmercury concentration in the liver of dead animals was 55.6 mg/kg ww. The mink were also fed a diet containing 10 mg/kg ww of mercuric chloride. The inorganic mercury had no observable effect on adult survival and reproduction even after five months of exposure. The average liver residue in these animals was 3.16 mg/kg ww.

Wobeser, Nielsen, and Schiefer (1976a) reported that a mercury concentration of 0.44 mg/kg ww in fish fed to a mink for 145 days had no observable adverse effects.

In another study, Wobeser, Nielsen, and Schiefer (1976b) fed mink a diet spiked with 1.1, 1.8, 4.8, 8.3, and 15.0 mg/kg ww of methylmercuric chloride for 93 days. All exposed animals developed symptoms of mercury poisoning. The onset of intoxication was negatively correlated with dose. One to two days were sufficient for the mink to develop mercury poisoning symptoms after exposure to 15.0 mg/kg. At least 70 days were needed to elicit the same response in animals exposed to 1.8 mg/kg ww. Some mink died as a result of the exposure. The dead animals contained mean mercury concentrations of 24.3, 23.1, and 11.9 mg/kg ww in the liver, kidney, and brain, respectively.

Jernelov *et al.* (1976) fed mink with fish that contained 5.7 mg/kg ww of methylmercury (5.8 mg/kg of total mercury). After 100 days of exposure, the mink appeared to be healthy and free from any symptoms of poisoning. The liver concentration at this point was 65 mg/kg ww, and the kidney concentration was 46.5 mg/kg ww. The apparent tolerance of an otherwise sensitive animal was attributed to the antagonistic action of selenium, which was also detected in the kidney and the liver.

In a two-generation study on mink by Dansereau *et al.* (1999), diets of 0.1 and 0.5 mg/kg total (T) Hg had no effect on survival. However, a dose of 1.0 mg/kg T Hg killed 30 of 50 first generation mink. Neurological and clinical signs of poisoning were seen between 75 and 100 days of exposure. In the second generation, six of seven mink in the 1.0 mg/kg T Hg died after 330 days of exposure. Death usually occurred seven days after the first clinical signs. Effects included weight loss, lethargy, incoordination, and splaying of hind legs (Dansereau *et al.* 1999).

5.4.3.3 Cat (*Felis domesticus*) and Florida Panther (*Puma concolor coryi*)

Takeuchi *et al.* (1977) fed a cat a diet composed of fish parts collected from a contaminated river in Ontario (English River). The cat became agitated and begun to jump into the air, growl, and froth at the mouth. Eventually, it became ataxic and anorexic. The dissection of the dead cat revealed severe deterioration of the brain cells, especially those of the cerebellum cortex. The liver of this animal contained 67.1 mg/kg ww of mercury. The brain and the kidney contained 16.4 and 13.4 mg/kg ww, respectively.

In an earlier study, Takeuchi *et al.* (1962) analyzed cats from Minamata Bay, Japan, that showed signs of mercury poisoning. The brain tissues contained 8.1 to 18.6 mg/kg of mercury.

Charbonneau *et al.* (1974) fed cats a diet composed of mercury-contaminated fish. The fish contained about 0.25 mg/kg of methylmercury. The cats developed signs of mercury poisoning after 55 days of exposure.

Albanus *et al.* (1972) fed cats a diet consisting of fish homogenate containing a mean methylmercury concentration of 6.0 mg/kg ww. The daily dose of methylmercury was about 0.45 mg/kg ww. The cats developed behavioral and locomotor abnormalities after 50 days of exposure. Convulsions began soon after. The dead cats were analyzed for mercury residues and the mean concentrations of methylmercury were 39, 31, and 18 mg/kg ww in the liver, kidney, and brain, respectively.

Eaton, Secord, and Hewitt (1980) fed cats a diet composed of liver from ringed seals that was contaminated with mercury to a level of 26.2 mg/kg ww, though only 3% of the total mercury was present as methylmercury. The methylmercury concentration was 0.79 mg/kg ww. The feeding regime was 25, 50, and 100 g of seal liver per day. Another regime included a diet of beef liver spiked with methylmercury chloride to give a dose of 0.25 mg of methylmercury per kg per day. A 90-day regime of seal liver produced no adverse effects. However, the beef liver regime caused convulsions after 68 days of exposure. The mean survival time for the latter group was 78 days. The analysis of livers of cats fed seal diet revealed mercury residue levels of 1.08, 2.27, and 3.66 mg/kg ww for the 25, 50, and 100 g treatments, respectively. The cats fed the beef diet had mean liver concentrations of 40.3 mg/kg ww.

Roelke *et al.* (1991) investigated mercury poisoning in Florida panther. An individual panther found dead (presumably due to mercury poisoning) contained 110 mg/kg ww in its liver. Blood levels of more than 0.5 mg/kg in females are associated with reduced kitten survival.

5.4.3.4 Ferrets (*Mustela nigripes*)

Hanko *et al.* (1970) fed ferrets a diet consisting of mercury-contaminated chickens. The dietary concentrations were 5 or 7 mg/kg ww. After two weeks, the higher dose caused the animals to feed less and exhibit signs of mercury poisoning. Death followed after 35 to 36 days of exposure. The lower dose required three weeks to produce the same initial response and death followed 58 days of exposure. The dead animals contained 53.7 and 69.0 mg/kg ww in their livers and kidneys, respectively.

5.4.3.5 Red Fox (*Vulpes vulpes*)

Borg *et al.* (1969) analyzed the liver of a fox that was behaving erratically and showing symptoms of mercury poisoning. The liver-kidney homogenate contained 30 mg/kg ww of mercury.

5.4.3.6 Marten (*Martes americana*)

Borg *et al.* (1969) also analyzed the liver of a marten behaving in a similar manner. The liver-kidney homogenate contained 40 mg/kg ww of mercury.

5.4.4 Summary of Effect Levels in Mammals

The review of mercury toxicity data for mammals suggests that a mercury residue of30 mg/kg ww in the liver or kidney is harmful and lethal to most mammals. Also, dietary methylmercury doses in the range of 2 to 6 mg/kg/day ww are considered sufficient to cause mercury poisoning.

5.5 Factors That Modify Toxicity

The toxicity of mercury to aquatic life and terrestrial wildlife is controlled by several physicochemical and biological factors. Some of these factors, such as presence of selenium, have a direct influence, where others, such as dissolved organic carbon, control the toxicity indirectly by modifying bioavailability. The factors are often interrelated, and it is difficult to separate the effect of any single factor.

The modifying factors introduce variability to the levels of mercury associated with toxic responses in biota. While it may not be possible to take into account all these factors, several major trends have emerged and are discussed below.

5.5.1 Mercury Speciation

Methylmercury is far more toxic than inorganic mercury. Methylmercury also bioaccumulates to a much greater extent than inorganic mercury. Therefore, it is important to determine which form of mercury is being analyzed for and the adverse effects that are associated with it.

5.5.2 Selenium

The protective effects of dietary selenium are well known. Ganther *et al.* (1972) described the antagonistic behavior of selenium against methylmercury toxicity to birds. Selenium exerts its protective action by binding mercury into mercury-selenoprotein complexes and making it unavailable to the sites of action (Palmisano, Cardellicchio, and Zambonin 1995; Cavalli and Cardelicchio 1995).

5.5.3 Dissolved Organic Carbon (DOC)

Dissolved organic carbon also has the potential to bind mercury and make it unavailable for uptake (Driscoll *et al.* 1994; 1995). Thus, high levels of DOC in ambient water might reduce mercury toxicity to aquatic life.

5.5.4 pH

Reduced pH stimulates the methylation of mercury in sediments (Spry and Wiener 1991) as well as biogeochemical releases (Rada, Powell, and Wiener 1993). The increased methylation rate will lead to higher levels of the toxic and bioaccumulative methylmercury (Xun, Campbell, and Rudd 1987; Bloom, Watras, and Hurley 1991). Lower pH also leads to lower levels of calcium, which has the ability to sorb mercury and make it unavailable for uptake.

5.5.5 Mixture Toxicity

In a natural setting, mercury often co-occurs with other contaminants. Thus, there is a possibility that other chemicals enhance or reduce the toxicity of mercury. Evidence exists that there is an additive or even a synergistic relationship between mercury and chlorinated compounds such as PCBs.

5.5.6 Temperature

Evidence suggests that temperature affects mercury toxicity to aquatic biota. U.S. EPA (1985) data indicated that mercury toxicity to fish increases with temperature. Temperature also controls the metabolic rate of bacteria that produce methylmercury. Thus, higher temperatures promote higher production levels of methylmercury and increase the potential for exposure and uptake of toxic methylmercury. Higher temperatures were negatively correlated with demethylation rates in lakes (Bodaly *et al.* 1993).

5.5.7 Metabolism

Mammals and birds of prey have the ability to demethylate methylmercury in the liver and kidneys (Norheim and Froslic 1978; Uthe, Solomon, and Grift 1972; Freeman and Horne 1973; Wren *et al.* 1980; Eaton, Secord, and Hewitt 1980; Jernelov *et al.* 1976). The organisms that can metabolize methylmercury include predatory birds such as the eagles and ospreys, marine mammals and birds, raccoons, otters, cats, and mink.

5.5.8 Vitamin E and Zinc

Vitamin E delays the onset of methylmercury toxicity in mammals (Welsh and Soares 1976) and protects against genotoxicity (Gilbert *et al.* 1983). Addition of zinc to a diet containing methylmercury caused rats to survive longer. Presumably, the addition of zinc caused the production of extra metalothioneins, which sequestered some mercury and kept it away from the sites of action.

CHAPTER 6.0

Field Research to Develop Ambient Water Quality Criteria Model

A field research program was undertaken to collect a dataset suitable for developing an ambient water quality criteria (AWQC) model for the protection of aquatic life and wildlife. A complete dataset with many measured variables (Table 6.1) collected from a large number of lakes was required for this purpose.

Table 6-1. Summary of Variables from Field Study for Use in Developing Mercury AWQC.

Parameter	Notes
Fish tissue ♦ Total mercury ♦ Methylmercury ♦ Length ♦ Weight ♦ Species	♦ Whole fish tissue samples or dorsal muscle samples from a minimum of 20 fish specimens (≤ 20 cm length). Whole fish samples preferred.
Lake surface water ♦ Total mercury ♦ Methylmercury ♦ Filtered ♦ Unfiltered ♦ DOC ♦ TOC ♦ Dissolved oxygen ♦ pH ♦ Conductivity ♦ Sulfide ♦ Sulfate ♦ Phosphorous ♦ Turbidity ♦ Colour ♦ Temperature ♦ Suspended solids	♦ Samples taken at approximately 0.5 m depth beyond the littoral zone of each lake. ♦ Samples analyzed for total and methylmercury. ♦ Samples filtered and unfiltered to determine the contribution of particulate-bound mercury ♦ Only lakes not receiving industrial discharge or municipal effluent discharge sampled

The scientific literature contains some of the data required to develop the AWQC model (Benoit *et al.* 1998; Gilmour *et al.* 1998; Sorensen *et al.*, 1990); however, these studies do not report all the necessary variables. In particular, surveys of fish mercury contamination, with the exception of Sorensen *et al.* (1990), have generally not reported any measure of mercury (total or methyl) concentration in lake water. This is generally because of the rigorous and expensive analytical methods needed to accurately measure total and methylmercury in water and because water mercury concentration is an insignificant predictor (a statistically insignificant regressor) of fish mercury concentration when combined in multiple regression analysis with other variables, particularly DOC concentration. In the case of Sorensen *et al.* (1990), data on total mercury concentration in lake water were not reported for individual lakes, thereby limiting possible manipulation and extrapolation of results from that study.

This section of the report presents the research conducted to specifically relate mercury levels in fish to total mercury and methylmercury in the water of lakes from which the fish were collected. Second, these data are used along with other important variables to develop a multiple regression model that can be employed to predict lake-specific mercury concentrations in fish. Concentrations of methylmercury in water, total mercury in water, and total mercury in small fish (4–21 cm) are reported from 42 lakes in Ontario and Nova Scotia. The mercury content of common species in each lake were collected and determined, because the goal was to develop a model that could be applied to a wide range of lakes, some of which might contain different prey species.

6.1 Methods

6.1.1 Study Lakes

From a database compiled by G. M. Richardson (RiskLogic Environmental Services Inc., pers comm. February 1st, 2002) that contained limnological data for more than 12,000 Ontario Lakes, 31 lakes were selected from the Eastern Ontario region that represent a range of water quality characteristics (DOC, pH, alkalinity) known to influence mercury bioavailability. Lakes with known point sources of anthropogenic mercury loadings (industries, sewage treatment plants, etc.) were excluded. Because the pH of Ontario lakes was comparatively narrow (~7–9), data were also obtained for lower pH lakes from Kejimkujik National Park, Nova Scotia, Canada, where mercury concentrations in loons are among the highest reported (Evers *et al.* 1998) and which have been the focus of recent mercury investigations (summarised by O'Driscoll *et al.* 2001).

6.1.2 Lake Water Quality

In Ontario lakes, water temperature, dissolved oxygen, specific conductance, and pH were measured in the top one meter of the lake during one sampling event. No accounting was made for the annual variability of DOC, pH, MeHg, or other parameters. All water parameters as well as Secchi depths were taken at least 50 m from shore. Surface (top 0.5 m) water samples were collected in pre-cleaned containers, transported on ice to the lab, and refrigerated at 4°C until analysis for DOC, sulfate, pH/alkalinity, total phosphorous, and true colour. Sulfate and total phosphorous were measured using standard spectrophotometric methods at the St. Lawrence River Institute laboratory (ATSDR 1997). Samples for DOC and true colour were first filtered through combusted GFF filters. True colour was measured spectrophotometrically at 455 nm, and DOC was measured by UV/ persulphate oxidation. Alkalinity was determined by potentiometric titration (ATSDR 1997). Water quality measurements of Nova Scotia lakes were determined using similar methods (O'Driscoll *et al.* 2001). Secchi disk depth and total phosphorous were not measured in the Nova Scotia lakes.

Water samples for mercury analysis were collected using care to minimize the potential for contamination of the samples. The samples were collected at each lake from the bow of a plastic-bottomed canoe positioned into the wind. Duplicate water samples were collected wearing elbow-length polyethylene gloves by immersing sealed and pre-cleaned polypropylene centrifuge tubes (total mercury samples) or 1-L Nalgene high density polyethylene bottles (methylmercury samples) beneath the water surface, uncapping the container, collecting the water sample and sealing the container under water. On shore, the samples were preserved with BrCl (total mercury samples) or trace metal quality HCl (MeHg) and double-bagged before being transported on ice to the University of Ottawa where they were analyzed.

Methylmercury analysis was performed using ultra-trace technique by gas chromatography/AFS following pre-concentration of the sample onto sulfhydril columns and eluting with solvent (Cai *et al.* 1996). Total mercury was analyzed according to the procedures outlined by Cai *et al.* (1996). Travel and procedural blanks were analyzed for every 10 samples, and QA/QC criteria followed EPA guidelines (U.S. EPA 2001). Method detection limits were 0.020 ng/L for MeHg and 0.5 ng/L for total mercury.

6.1.3 Fish Collection

Using seines and angling, approximately 20 to 100 fish were collected from each Ontario lake, representing at least two or three common species per lake. Each fish was identified, measured (total length), and weighed before being placed in an individual zippered plastic bag with a unique identification number. Fish were transported on ice to the laboratory, where they were frozen at –20°C. Fish remained frozen until they were prepared for mercury analysis.

Once the field collections were completed, a subset of fish, typically 20 specimens representing two to three species per lake, were selected for mercury analysis. The individuals and species chosen for mercury analysis were selected to provide similar species and size ranges of fish from each lake. Whenever possible, fish ranging in size from 5–20 cm were selected because this is the size range likely to be consumed by mink, the most sensitive wildlife species with respect to methylmercury toxicity (Moore *et al.* 1999a). Mink are opportunistic foragers (U.S. EPA 1993c), and, when feeding on fish, are not species selective. Thus, the sampling protocol mimicked the potential prey selection of mink that consume small fish from littoral zones. This selection procedure introduced variability associated with inter-specific differences in the mercury content of prey. However, it was essential to include this variability in the data to develop a model that would be applicable across a wide geographic range. Analyses of inter-specific differences in mercury content of fish will be the topic of another paper in this series.

Nova Scotia fish were collected primarily with minnow traps. Between two and 30 fish, representing common species, were captured from each Nova Scotia lake. In most cases, because fewer fish were obtained from Nova Scotia Lakes, all individuals were analyzed for mercury content.

6.1.4 Total Mercury in Fish

Fish were thawed and homogenized in a food processor until a uniform consistency was obtained. A sub-sample (approximately 0.5 g) of this homogenate was used for mercury analysis, and the remainder was re-frozen and archived for future reference. A sample of dorsal muscle was removed in a subset of fish prior to homogenization, and the dorsal muscle and whole fish samples were both analyzed. The fish samples were weighed into pre-cleaned digestion tubes and digested by hot H_2SO_4:HNO_3, followed by $KmnO_4/H_2O_2$ oxidation, and analyzed by cold vapor atomic absorption spectrophotometry with a Varian CETAC M-6000A mercury analyzer. Each run of 48 fish samples included standards, replicate samples (10% of fish samples), duplicate matrix spikes, triplicate method blanks, and duplicate analyses of standard reference material (Dorm-2, National Research Council, Ottawa, ON). All equipment was acid cleaned and thoroughly rinsed with distilled water between samples.

6.2 Statistical Analysis

All statistical analyses were conducted using SYSTAT version 5.03 (1993) statistical software. Data were log transformed when necessary to meet the assumptions of statistical tests (normality, linearity, and homoscedasticity). Simple regression and correlation analysis was used to examine the relationship of water chemistry variables to one another and to mercury (methyl and total) in water and fish (total). To ensure that the inter-lake variability caused by differences in fish length was controlled prior to evaluating the role of water chemistry on fish mercury accumulation, the concentration of mercury in fish of a standard 10-cm total length was determined on a lake-by-lake basis from the regression of $\log_{10}$ fish mercury concentration on $\log_{10}$ fish total length. Further analysis by simple and multiple regression was conducted using the $\log_{10}$ 10-cm fish [Hg] as the dependent variable and lake water chemistry variables as independent variables.

6.2.1 Regression Model

Multiple regression analyses of $\log_{10}$ 10-cm fish [Hg] on each and every combination of water chemistry variables were performed. The final multiple regression equation predicting $\log_{10}$ 10-cm fish [Hg] from water chemistry variables was the best combination of significant independent variables ($p \leq 0.05$) that produced the highest r^2 value.

6.3 Results

6.3.1 Water Quality

The concentration of total mercury in water from Ontario lakes (n = 31) varied from 0.55 to 2.1 ng/L. The concentration of methylmercury in water from Ontario lakes ranged from 0.02 to 0.18 ng/L. In Nova Scotia, the concentration of mercury in lake water varied from 0.09 to 7.4 ng/L and 0.028 to 0.255 ng/L for total (n = 11 lakes) and methylmercury (n = 9 lakes), respectively.

As shown in Figures 6-1 and 6-2, $\log_{10}$ total mercury in water was correlated with $\log_{10}$ methylmercury in water ($r = 0.503$, $p = 0.001$, $n = 40$), pH ($r = -0.807$, $p < 0.001$, $n = 42$), $\log_{10}$ conductivity ($r = -0.696$, $p < 0.001$, $n = 42$), $\log_{10}$ color ($r = 0.816$, $p < 0.001$, $n = 42$) and sulphate ($r = -0.34$, $p = 0.032$, $n = 42$). Log_{10} methylmercury in water was correlated with pH ($r = -0.357$, $p = 0.024$, $n = 40$) and $\log_{10}$ color ($r = 0.476$, $p = 0.002$, $n = 40$).

Other water quality measurements were obtained from all 31 Ontario lakes (Table 6-2), with the exception of alkalinity that was not measured in one lake (Christie Lake). Ranges of water chemistry parameters were as follows: pH 7.1–9.4, dissolved oxygen 7.1–14.0 mg/L, sulphate ND-21 mg/L, DOC 3.4–13.8 mg C/L, true color 2.0–18.0 Hazens, Secchi disk depth 114–669 cm, alkalinity 8–192 mg/L, specific conductivity 43.3–361.1 μS/cm, and total phosphorous 6.6–41.1 μg/L.

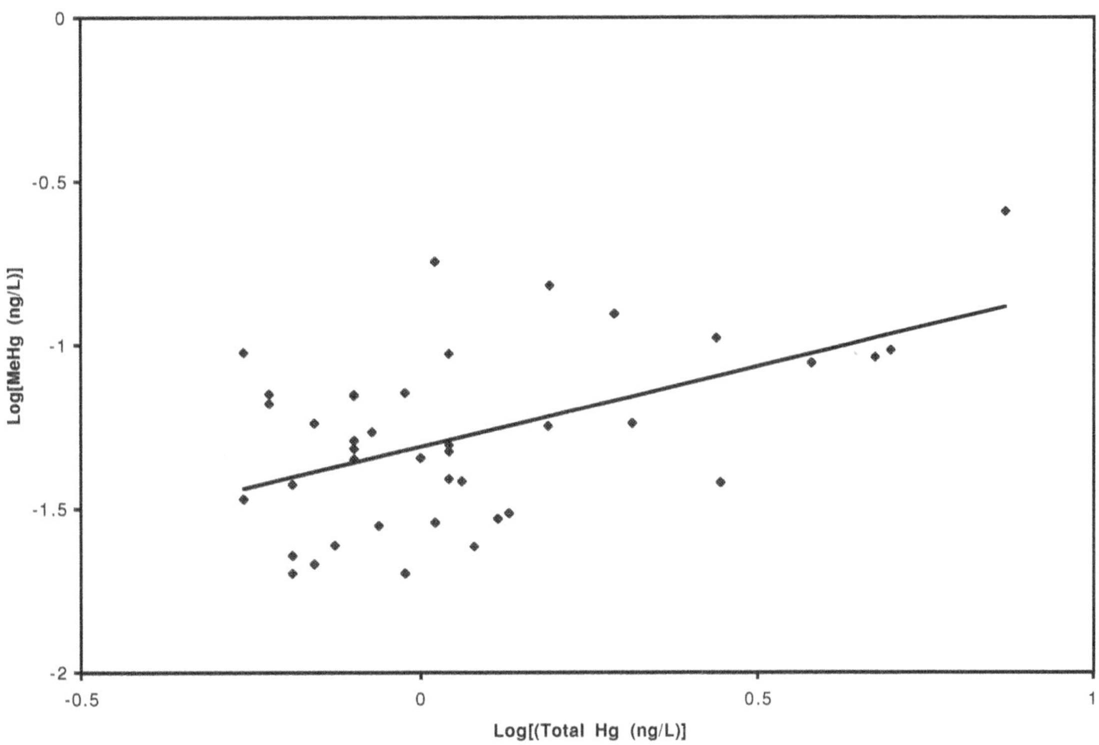

Figure 6-1. Relationship Between Total and Methylmercury in Water for 40 Study Lakes Across Two Provinces.

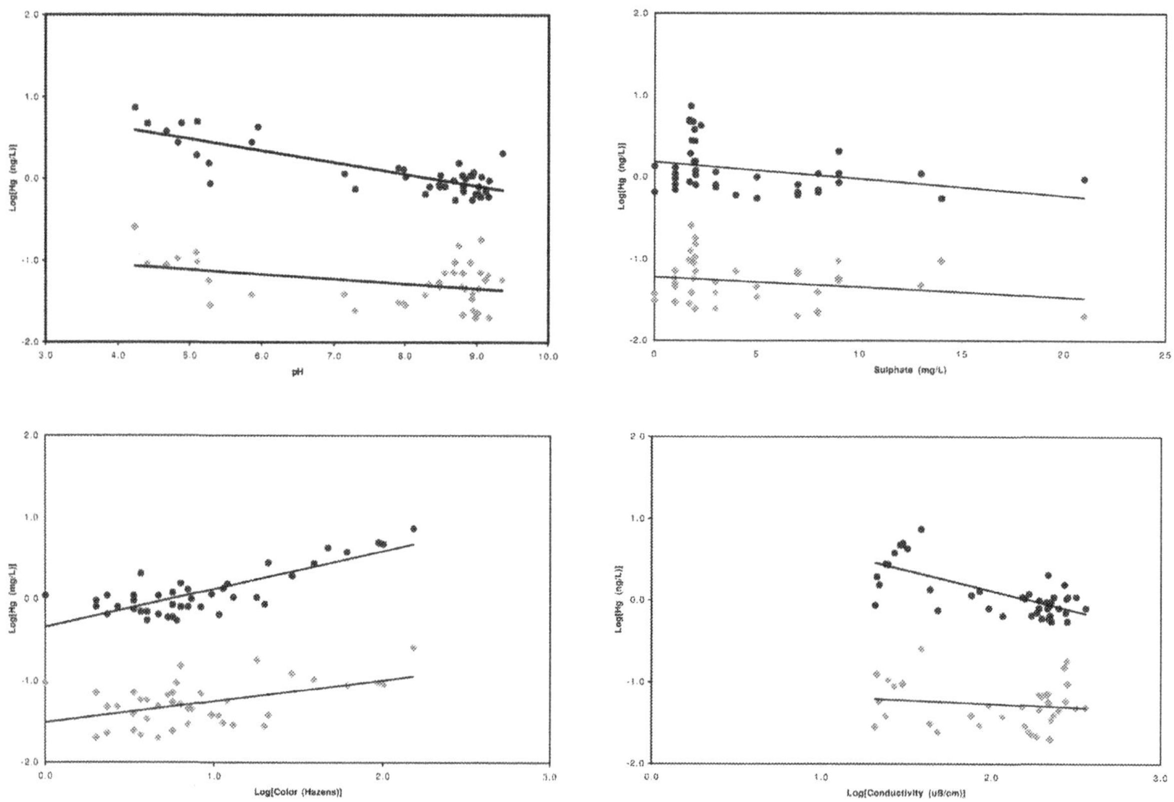

Figure 6-2. Relationships Between Aqueous Total Hg Concentrations and Aqueous Methylmercury Concentrations and Four Important Water Chemistry Parameters.
Note: Dark diamonds, n = 42 and Light-colored diamonds, n = 40

Nova Scotia lakes were generally lower in pH and alkalinity and higher in true color (Table 6-2). Ranges of water chemistry parameters for the Nova Scotia Lakes (n = 11) were: pH 4.2–5.9, sulphate 1.7–2.3 mg/L, DOC 2.2–13.1 mg C/L, true color 12–155 Hazens, alkalinity –1.27–1.91 mg/L, specific conductivity 21.1–38.5 μS/cm.

For all lakes combined, alkalinity was correlated with conductivity ($r = 0.974$, $p < 0.001$, $n = 41$). Log_{10} DOC was positively correlated with pH ($r = 0.448$, $p = 0.003$, $n = 42$), and log_{10} conductivity ($r = 0.594$, $p < 0.001$, $n = 42$). Log_{10} conductivity was positively correlated with pH ($r = 0.927$, $p < 0.001$, $n = 42$) and log_{10} color ($r = -0.778$, $p < 0.001$, $n = 42$). Sulphate was correlated with log_{10} conductivity ($r = 0.449$, $p = 0.003$, $n = 42$) and log_{10} color ($r = -0.502$, $p < 0.001$, $n = 42$).

6.4 Fish Collected

Total mercury measurements were obtained from 747 fish representing seven species (Table 6-3). Yellow perch (*Perca flavescens*), pumpkinseed sunfish (*Lepomis gibbosus*), and bluegill sunfish (*Lepomis macrochirus*) were the most frequently captured fish, being caught in 34, 30, and 19 lakes, respectively. Other species included common shiners (*Notropis cornutus*, two lakes), golden shiners (*Notemigonus crysoleucas*, two lakes), bluntnose minnow (*Pimephales notatus*, one lake), and banded killifish (*Fundulus diaphanus*, four lakes).

Table 6-2. Lake Location and Water Chemistry Parameters for 31 Lakes in Ontario and 10 Lakes in Nova Scotia, Canada.

Lake Name	Latitude	Longitude	pH	DO mg/L	SO_4 mg/L	DOC mg C/L	True Color Hazens	SECCHI Cm	ALK. Mg/L	Spec. Cond. uS/cm	Total Hg ng/L	MeHg ng/L	TP ug/L
ONTARIO													
Ashby	45.083N	77.350W	7.3	9.6	3.0	5.1	3.3	669	8	48.1	0.8	0.024	6.6
Beaver	44.500N	77.033W	8.0	7.1	1.0	7.6	13.0	284	62	157.0	1.1	0.029	12.3
Bellamys	44.717N	76.033W	9.1	9.1	1.0	9.9	4.0	161	125	273.5	0.7	0.058	13.8
Big Rideau	44.750N	76.217W	9.4	9.9	9.0	3.4	3.7	510	88	214.5	2.1	0.058	9.5
Buckshot	45.000N	77.067W	8.0	8.2	1.0	6.2	7.0	590	32	84.6	1.3	0.029	7.3
Calabogie	45.267N	76.750W	8.3	9.4	ND	6.2	10.7	241	44	116.6	0.7	0.037	11.1
Camden	44.417N	76.867W	8.8	9.6	2.0	13.8	6.3	118	112	269.3	1.6	0.151	24.4
Charleston	44.533N	76.000W	8.9	9.3	8.0	5.4	3.3	368	103	232.7	1.1	0.039	13.0
Christie	44.800N	76.433W	8.5	8.3	1.0	8.8	4.7	510	.	151.2	1.1	0.049	10.9
Crotch	44.917N	76.800W	8.4	8.0	3.0	5.9	6.3	526	39	96.4	0.8	0.051	13.7
Devil	44.583N	76.450W	8.8	8.3	8.0	4.6	3.7	575	81	184.2	0.7	0.021	7.3
Dog	44.417N	76.350W	9.0	11.5	8.0	5.8	2.3	328	66	171.0	0.7	0.023	19.7
Eloida	44.667N	75.967W	8.9	8.7	9.0	8.5	1.0	265	127	282.3	1.1	0.094	13.0
Gananoque	44.567N	75.883W	9.0	12.2	1.0	7.1	7.0	260	109	248.3	0.8	0.045	21.6
Graham	44.450N	76.150W	8.6	9.8	2.0	7.8	8.3	188	78	189.8	0.8	0.070	18.0
Indian	44.600N	76.333W	9.0	8.4	7.0	6.2	4.7	387	97	221.2	0.7	0.020	13.2
Lime	44.400N	77.117W	8.5	8.0	1.0	5.2	2.7	197	192	361.1	0.8	0.048	13.0
Little Mellon	44.350N	77.100W	7.1	7.5	3.0	6.2	9.7	480	21	75.9	1.2	0.038	10.2
Loch Garry	45.250N	76.700W	8.7	9.3	1.0	11.5	3.3	114	99	209.2	1.0	0.071	20.1
Loughborough	44.383N	76.533W	8.7	14.0	14.0	6.0	6.0	385	92	280.0	0.6	0.095	18.0
Mink	45.550N	77.050W	8.8	8.3	13.0	8.5	2.3	285	120	316.1	1.1	0.047	10.6
Newboro	44.633N	76.333W	9.1	9.4	4.0	6.7	5.7	197	87	216.3	0.6	0.071	16.8
Norway	45.333N	76.717W	8.9	9.3	5.0	8.6	4.0	463	110	225.2	0.6	0.034	6.6
Odessa	44.317N	76.683W	9.1	9.2	2.0	13.1	18.0	153	87	277.3	1.1	0.180	41.1
Opinicon	44.567N	76.317W	8.5	10.9	9.0	6.8	5.7	360	90	219.0	0.9	0.054	19.4
Pike	44.783N	76.350W	9.0	11.6	2.0	10.6	5.7	232	39	165.4	1.2	0.024	18.0

Table 6-2. Lake Location and Water Chemistry Parameters for 31 Lakes in Ontario and 10 Lakes in Nova Scotia, Canada (continued).

Lake Name	Latitude	Longitude	pH	DO mg/L	SO_4 mg/L	DOC mg C/L	True Color Hazens	SECCHI Cm	ALK. Mg/L	Spec. Cond. uS/cm	Total Hg ng/L	MeHg ng/L	TP ug/L
Sand	44.567N	76.267W	8.8	7.7	5.0	7.0	7.3	357	75	190.5	1.0	0.045	19.4
Troy	44.517N	76.267W	9.2	8.1	21.0	6.5	2.0	119	76	223.9	1.0	0.020	26.6
Varty	44.383N	76.817W	8.8	7.8	7.0	7.1	2.0	154	85	213.7	0.8	0.070	14.5
Weslemkoon	45.033N	77.417W	7.9	7.1	ND	5.8	11.3	430	10	43.3	1.4	0.031	8.0
Whitefish	44.533N	76.233W	9.2	8.4	7.0	5.5	5.3	336	80	196.9	0.6	0.066	15.1
Nova Scotia													
Beaverskin	49.085N	31.393W	5.3	.	1.9	2.2	12	.	0.92	21.8	1.5	0.057	.
Big Dam East	49.243N	31.976W	5.9	.	1.8	2.9	21	.	1.91	23.8	2.8	0.038	.
Big Dam West	49.255N	31.783W	5.1	.	1.7	5.6	94	.	0.54	30.1	5.0	0.096	.
Big Red	49.131N	31.032W	4.2	.	1.8	13.1	155	.	-1.27	38.5	7.4	0.255	.
Frozen Ocean	49.245N	31.302W	4.9	.	1.7	5.6	97	.	-0.14	29.0	4.8	.	.
Grafton	49.170N	32.667W	5.9	.	2.3	4.1	47	.	1.31	32.0	4.3	.	.
North Cranberry	49.113N	32.192W	5.1	.	1.7	2.9	29	.	-0.06	21.1	1.9	0.125	.
Pebbleloggitch	49.080N	31.249W	4.4	.	1.9	7.1	101	.	-0.64	30.1	4.8	0.092	.
Peskawa	49.100N	31.172W	4.7	.	1.9	4.1	61	.	-0.39	26.9	3.8	0.088	.
Peskowesk	49.093N	31.770W	4.8	.	2.0	3.6	39	.	-0.30	24.7	2.7	0.105	.
Puzzle	49.102N	32.215W	5.3	.	1.7	2.9	20	.	0.55	20.6	0.9	0.028	.

Table 6-3. Mean Fish Length, Width, and Number of Fish Captured for Mercury Analysis From 31 Lakes in Ontario and 11 in Nova Scotia, Canada.

Lake Name	Bluegill			Pumpkinseed			Perch			Other	
	N	Mean Length	Mean Mass	N	Mean Length	Mean Mass	N	Mean Length	Mean Mass	N	Species
Ontario											
Ashby	.	.	.	10	10.9	22.6	10	12.4	27.5	.	.
Beaver	6	7.6	10.5	5	12.5	39.1	6	10.3	13.5	.	.
Bellamys	7	6.2	3.6	.	.	.	6	10.2	11.6	7	BNM
Big Rideau	9	11.9	35.4	8	11.6	31.0	4	15.5	42.5	.	.
Buckshot	.	.	.	15	10.8	23.2	.	.	.	5	CS
Calabogie	.	.	.	10	11.2	27.4	10	11.6	19.1	.	.
Camden	.	.	.	12	13.5	52.9	8	13.6	29.6	.	.
Charleston	4	13.8	47.9	16	10.3	24.5	.	.	.	.	.
Christie	4	10.8	33.5	7	11.0	27.9	9	14.2	36.2	.	.
Crotch	.	.	.	2	10.0	16.4	12	5.4	1.8	6	GS
Devil	7	10.6	20.1	8	11.2	28.7	5	15.6	54.9	.	.
Dog	10	8.9	12.8	9	10.5	22.9	1	16.4	52.8	.	.
Eloida	.	.	.	16	13.1	57.9	4	8.7	10.2	.	.
Gananoque	7	13.2	51.1	7	11.4	33.2	6	18.1	69.2	.	.
Graham	9	8.0	9.0	2	10.3	17.4	8	9.9	11.4	1	GS
Indian	11	14.8	56.7	9	13.9	49.4	1	13.9	29.6	.	.
Lime	.	.	.	9	12.5	38.5	11	11.1	14.3	.	.
Little Mellon	8	8.8	10.8	8	8.2	9.8	4	7.6	4.6	.	.
Loch Garry	.	.	.	10	12.8	43.3	10	13.5	26.4	.	.
Loughborough	10	11.9	33.5	10	10.6	23.0	.	.	.	.	.
Mink	.	.	.	20	11.2	29.3	.	.	.	.	.
Newboro	10	11.0	27.8	10	11.0	27.8	.	.	.	.	.
Norway	7	8.4	10.9	4	14.7	65.1	9	10.1	12.1	.	.
Odessa	.	.	.	10	10.9	28.0	10	12.4	23.3	.	.
Opinicon	7	13.8	61.7	10	13.9	68.6	3	10.2	12.5	.	.
Pike	4	5.5	2.2	7	10.3	27.2	9	13.3	32.7	.	.
Sand	7	11.0	25.2	6	11.9	37.4	7	6.7	3.2	.	.

Table 6-3. Mean Fish Length, Width, and Number of Fish Captured for Mercury Analysis From 31 Lakes in Ontario and 11 in Nova Scotia, Canada (continued).

Lake Name	Bluegill			Pumpkinseed			Perch			Other	
	N	Mean Length	Mean Mass	N	Mean Length	Mean Mass	N	Mean Length	Mean Mass	N	Species
Troy	8	12.8	41.0	8	14.7	68.1	4	9.1	8.2	.	.
Varty	.	.	.	10	13.5	49.6	5	16.4	52.4	.	.
Weslemkoon	.	.	.	20	14.3	58.0	.	.	.	.	.
Whitefish	17	7.9	7.9	1	6.5	4.0	2	14.1	32.4	.	.
NOVA SCOTIA											
Beaverskin	.	.	.	.	.	.	.	.	.	20	GS & BK
Big Dam East	.	.	.	.	.	.	9	7.8	5.4	5	BK
Big Dam West	.	.	.	.	.	.	20	7.6	5.0	.	.
Big Red	.	.	.	.	.	.	10	5.8	2.5	.	.
Frozen Ocean	.	.	.	.	.	.	20	7.1	4.4	.	.
Grafton	.	.	.	.	.	.	7	10.3	10.0	5	BK
North Cranberry	.	.	.	.	.	.	4	10.7	13.2	.	.
Pebbleloggitch	.	.	.	.	.	.	7	6.8	3.5	.	.
Peskawa	.	.	.	.	.	.	.	.	.	2	CS
Peskowest	.	.	.	.	.	.	8	7.4	4.6	.	.
Puzzle	.	.	.	.	.	.	11	9.3	8.1	5	CS

Other species were banded killfish (BK), common shiner (CS), golden shiner (GS), and bluntnose minnow (BNM).

6.5 Mercury in Fish

6.5.1 QA/QC Results

Method blanks contributed an average of 1.1 ± 1.0 ng (n = 85) of mercury to samples ranging from 6 ng to 105 ng Hg. The method detection limit, defined as three times the standard deviation of the blank, is therefore 3.1 ng Hg, corresponding to 6.2 ng/g Hg for a 0.5-g sample. Overall, analyses of DORM-2 averaged 103 ± 9% (range 76% to 133%, n = 52) of the certified value. Matrix spike samples showed good recovery at 107 ± 19%, suggesting negligible bias from matrix effects.

6.5.2 Fish Concentrations

The concentration of total mercury in 615 fish from 31 Ontario lakes ranged from 9 to 212 ng/g ww. Concentrations of mercury in 132 fish from 11 Nova Scotia lakes varied from 64 to 355 ng/g ww. Fish from Nova Scotia had significantly higher mercury concentrations than fish from Ontario (ANOVA, $p < 0.001$; Figure 6-3). Total mercury concentration was positively correlated with fish length for both provinces (Nova Scotia $r = 0.56$, $p < 0.001$; Ontario $r = 0.28$, $p < 0.001$).

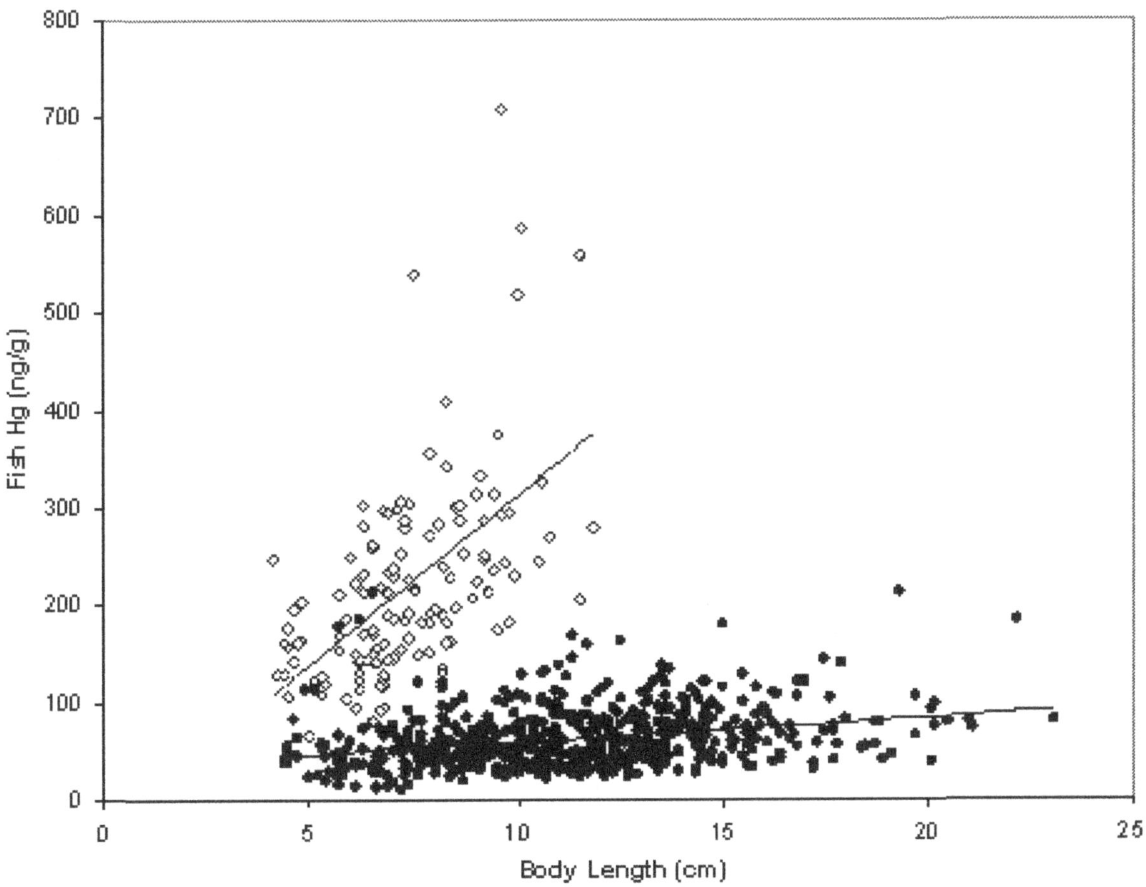

Figure 6-3. Comparison of all Fish Lengths and Total Hg in Fish (n = 747) for 31 Ontario Lakes (Filled Circles) and 11 Nova Scotia Lakes (Diamonds).

Much of the literature is based on the concentration of mercury in dorsal muscle (Haines *et al.* 1995). Therefore, to allow comparison, a subset of 59 fish were analyzed using both dorsal and whole fish (Figure 6-4). A regression between dorsal and whole fish mercury concentrations revealed a significant correlation ($r = 0.9161$, $p < 0.001$). The regression equation ($y = 1.3386 \times + 14.504$) can be used to predict dorsal muscle from whole fish concentration.

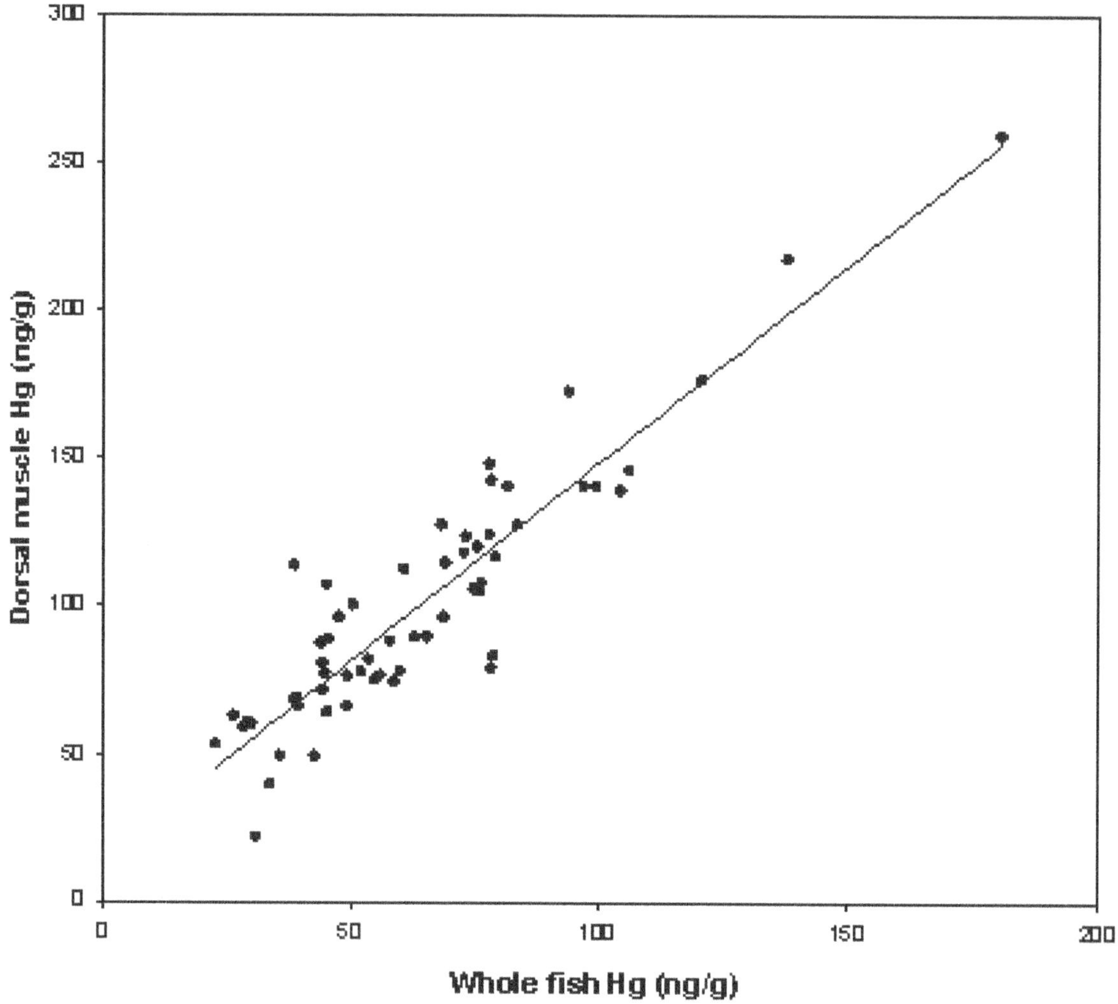

Figure 6-4. Comparison of Dorsal Muscle Hg Concentration and Whole Fish Hg Concentration (n = 59) from 26 Study Lakes.

Fish mercury concentration measured in this study was strongly correlated with fish length. The study controlled for the effect of fish length by calculating the mercury concentration of a standard 10-cm fish (approximate median size of prey consumed by mink). The regression of log length against log fish mercury concentration on a lake-by-lake basis was used to calculate the predicted mercury concentration of a standard 10-cm fish for each lake. These predicted fish mercury concentration values (for a 10-cm fish) were used in all further analyses with water quality variables. As shown in Figure 6-5, $\log_{10}$ total mercury of a standardized 10-cm fish was correlated with $\log_{10}$ total mercury in water ($r = 0.635$, $p < 0.001$, $n = 38$), $\log_{10}$ MeHg in water ($r = 0.339$, $p = 0.037$, $n = 38$), $\log_{10}$ DOC ($r = -0.413$, $p = 0.01$, $n = 38$), pH (r = –0.881, p < 0.001, n = 38), $\log_{10}$ color (r = 0.71, p < 0.001, n = 38), $\log_{10}$ conductivity (r = –0.788, p < 0.001, n = 38), and sulphate (r = –0.388, p = 0.016, n = 38).

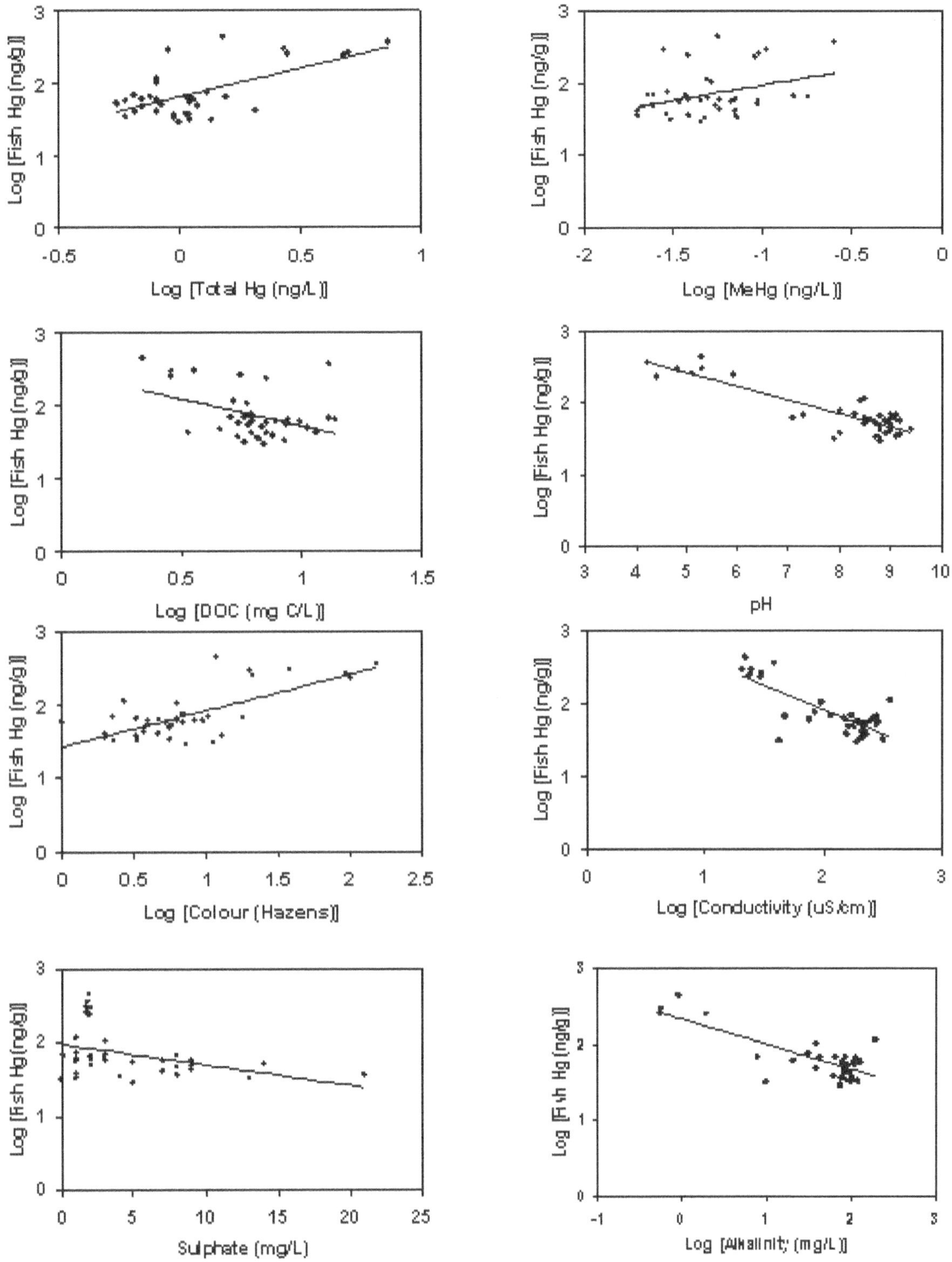

Figure 6-5. Fish Hg Concentrations for Standard Fish Length of 10 cm Compared to Significant Water Chemistry Parameters Across 38 Lakes.

As mentioned above, fish mercury concentration measured in this study was correlated with fish length. For the seven individual Nova Scotia lakes considered for further analysis, the correlation coefficients (r) of $\log_{10}$ fish [Hg] on $\log_{10}$ fish length ranged from 0.623 to 0.928. For the 31 Ontario lakes, individual lake r-values for the $\log_{10}$ fish [Hg] to $\log_{10}$ fish length relationship ranged from –0.628 to 0.893. Three Ontario lakes produced negative Hg-length correlations, because of the variable mix of fish species and the subsequent influence of interspecies differences in Hg accumulation on that regression. Irrespective of this result, the regression of log length against log fish mercury concentration on a lake-by-lake basis was used to calculate the predicted mercury concentration of a standard 10-cm fish for each lake. These predicted fish mercury concentration values (for a 10-cm fish, approximate median size of prey consumed by mink), were used in all further analyses with water quality variables.

As shown in Figure 6-5, $\log_{10}$ total mercury of a standardized 10-cm fish was correlated with $\log_{10}$ total mercury in water ($r = 0.635$, $p < 0.001$, $n = 38$), $\log_{10}$ MeHg in water ($r = 0.339$, $p = 0.037$, $n = 38$), $\log_{10}$ DOC ($r = -0.413$, $p = 0.01$, $n = 38$), pH ($r = -0.881$, $p < 0.001$, $n = 38$), $\log_{10}$ color ($r = 0.71$, $p < 0.001$, $n = 38$), $\log_{10}$ conductivity ($r = -0.788$, $p < 0.001$, $n = 38$), and sulphate ($r = -0.388$, $p = 0.016$, $n = 38$). Simple regression results by province (NS = 7 lakes; Ontario = 31 lakes) proved insignificant for all water chemistry variables, due largely to the narrow range of water chemistry values within the lakes from the individual provinces. Although the study lakes might, therefore, appear to suggest two disparate datasets, the data available from another study (Greenfield *et al.* 2001) demonstrate that these results fit nicely at two extremes of a continuum (discussed further below).

6.5.3 Multi-Regression Analysis

From the initial dataset (n = 42 lakes) two Nova Scotia lakes were eliminated for which there was no methylmercury data and two other Nova Scotia lakes with too few fish to calculate the Hg concentration of a standard 10-cm fish. The remaining 38 lakes, 31 from Ontario and seven from Nova Scotia were included in the multiple regression model. The best model to predict mercury concentration in a standard 10-cm fish from lake water chemistry variables was:

$$\text{Log}_{10}(10\text{ cm } C_f) = C - RC_{pH}\, x\ pH - RC_{DOC}\, x\, \text{Log}_{10}(DOC) + RC_{C_w}\, x\, \text{Log}_{10}(C_w) \qquad (6.5.3\text{-}1)$$

where C_f (µg/kg) is the concentration of mercury in fish, C (unitless) is a constant, RC_{pH} (unitless) is the regression coefficient for pH, pH (unitless) is a user-entered parameter for the water body of interest, RC_{DOC} (unitless) is the regression coefficient for dissolved organic carbon, DOC (mg C/L) is a user-entered parameter for the water body of interest, RC_{Cw} (unitless) is the regression coefficient for concentration of methylmercury in water, and C_w (pg/L) is the concentration of methylmercury in water. The above regression equation was standardized to 10-cm fish because this is the approximate median size for fish consumed by mink (U.S. EPA 1993c).

The stepwise multiple regression identified pH as the most important factor controlling mercury bioaccumulation in fish. The influence of pH alone explained 77.7% of the variation in mercury concentration in fish. Once variation in pH was accounted for, total mercury concentration in water failed to significantly increase the predictive power of the model. Methylmercury in water and DOC, in contrast, increased the amount of variation explained by the model to 80.4%. This increase, though small, was statistically significant (DOC $p < 0.023$; MeHg

$p < 0.03$). Examination of the interaction between DOC, pH, and MeHg indicated that there was no co-linearity between the independent variables ($p > 0.178$ for all possible combinations).

6.6 Discussion

This study found significant correlations between fish mercury concentrations and fish length, total Hg and MeHg in water, pH, DOC, colour, conductivity, sulphate, and alkalinity. In addition, a significant relationship was found between log total Hg in water and log MeHg in water.

Because pH was known to be an important factor in regulating mercury bioavailability (Cope, Wiener, and Rada 1990; Sorenson 1990) the study lakes were selected from two provinces. The lakes in Nova Scotia were known to be low in pH (O'Driscoll *et al.* 2001). When pH was plotted against mercury concentration (in fish or water), the two provinces clustered at opposite corners of the graph (Figure 6-6). Therefore, there was a concern that the two systems were not part of a continuous distribution with respect to pH. This possibility was investigated by obtaining data from the literature for lakes with pH in the range of 6–7. When pH was plotted against mercury concentration in fish and the data from Greenfield *et al.* (2001) along with this study's data was included, the distribution appeared continuous (Figure 6-6). Furthermore, the slope was almost identical to the slope in Greenfield *et al.* (2001).

The fish mercury data were calculated at the mercury concentration in a standard 10-cm fish. Because the mercury concentration in a 10-cm fish could not be calculated from the data presented in Greenfield *et al.* (2001), instead their averages were plotted along with this study's standard mercury concentrations. Although the mercury concentrations were derived in a different way, most of their fish were of similar size (9.7–21.1 cm). The results of this analysis did not change when only the fish from the Greenfield *et al.* (2001) study with mean length close to 10 cm were plotted. This relationship also held whether all the fish or just perch, the species examined by Greenfield *et al.* (2001), were plotted. A similar relationship was also found between pH and fish mercury concentration for similar size range of fish from northern Michigan lakes (Watras *et al.* 1998).

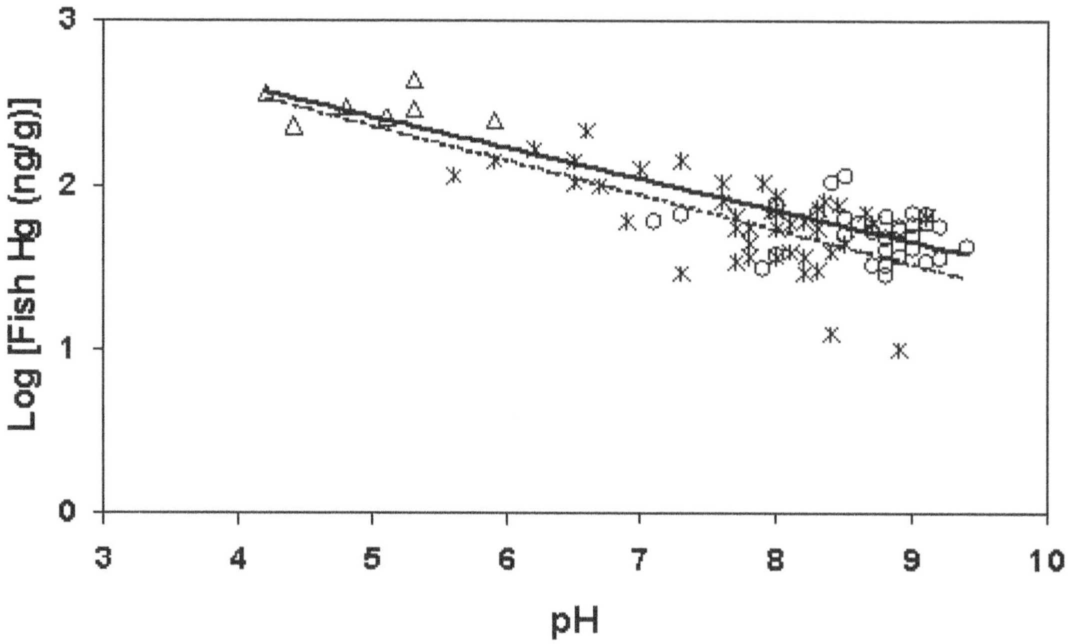

Figure 6-6. Relationship Between pH and Fish Hg Concentrations for Standard Fish Length of 10 cm for 38 Lakes with Average Fish Mercury Concentration Data from 42 Wisconsin Seepage Lakes.
Note: 31 Ontario lakes are shown as circles and 7 Nova Scotia lakes are shown as triangles.

The role of pH in fish mercury accumulation may relate to its influence on methylation rate (Gilmour and Henry 1991; Miskimmin, Rudd, and Kelly 1992; Xun *et al.* 1987) and also to its influence on DOC solubility, thereby indirectly affecting Hg bioavailability (Richardson and Currie 1995; reviewed by Richardson, 1994). Stein *et al.* (1996) summarized the possible mechanisms that have been brought forward for the increase in mercury methylation rate with decreased pH:

1. Increased oxidant levels in the atmosphere result in more oxidation of elemental mercury to soluble species that are subsequently deposited to surface waters. This increases the amount of mercury available for methylation.
2. Acidification increases the leaching of mercury from soil and increases the mercury in surface runoff or groundwater flow.
3. Low pH decreases the binding of mercury to humic substances and causes some humics to precipitate.
4. Low pH decreases the role of humics as electron donors in the reduction of Hg^{2+} to volatile Hg^0.
5. Reduced volatilization rates at low pH increase the availability of mercury for methylation.
6. Microbial mercury methylation is increased at the sediment-water interface in low pH waters, and the rate of microbial demethylation is decreased.
7. In low pH conditions, the formation of the soluble and bioavailable monomethylmercury is favored over the formation of the very volatile dimethylmercury species.

Cope, Wiener, and Rada (1990) found that pH explained about 64% of the variation in mercury concentration in whole perch from 10 Wisconsin seepage lakes. Greenfield *et al.* (2001) working with a larger set of Wisconsin lakes (seepage, drainage, and headwater lakes), found that pH was the most important factor in determining mercury concentration in one- to three-year-old yellow perch. The relationship between pH and mercury in fish seems to be robust and holds across a wide range of lake types and species. Decreased pH may increase methylmercury in water by increasing methylation of mercury at the sediment-water interface and possibly in the water column (Winfrey and Rudd 1990). Reduced pH may also increase the bioavailability of Hg to fish by decreasing the loss of volatile Hg from water and increasing Hg binding to particles in the water column (Winfrey and Rudd 1990), thereby making more Hg available for bioaccumulation. This study found that the concentration of methylmercury in water was positively correlated with pH, supporting the hypothesis that decreased pH increases the bioavailability of mercury.

In the sampled lakes, DOC was negatively correlated with the mercury concentration in fish, but color showed a relatively strong positive correlation with fish Hg. The relationship between DOC or color and mercury in fish is less clear. Driscoll *et al.* (1995) found that DOC was positively correlated with the concentration of mercury in yellow perch at DOC concentrations below 8 mg C/L. Above this concentration, the correlation between mercury in perch and DOC became negative. This study's DOC values ranged from about 2 to 14 mg C/L, a much smaller range than the lakes studied by Driscoll *et al.* (1995).

The influence of DOC on fish mercury likely relates to its correlation with mercury loading from watersheds into lakes (Lee and Iverfeldt 1991; Mierle and Ingram 1991; Richardson and Currie 1995), its role in complexation and bioavailability of Hg in the water column (Richardson and Currie 1995), and possibly abiotic methylation (Weber 1993). DOC can also influence the rate at which Hg(II) is reduced to elemental mercury. When water samples are exposed to solar radiation, there can be a decline in the rate of photoreduction with increased DOC (Amyot *et al.* 1997). In other words, DOC may act as a competitive inhibitor. Higher DOC loading increases the influx of mercury to lakes and reduces loss rates, resulting in greater mercury retention and accumulation in the water body. The conflicting role of DOC may result from its dual role with respect to mercury bioavailability. On one hand, the transport of humic material from wetlands and forests increases the influx of DOC-bound mercury. Yet, DOC in the water column may bind methylmercury, making it unavailable to aquatic organisms (Driscoll *et al.* 1995).

As previously noted, this study tested the three assumptions on which the EPA guidelines for the protection of wildlife are based: that the concentration of mercury in fish is a function of 1) the concentration of total mercury in lake water, 2) methylmercury levels in lake water, and that 3) total and methylmercury levels in lake water are related. Although concentrations of both methyl and total mercury in water were positively correlated with the concentration of mercury in a standard 10-cm fish, only methylmercury was retained as a significant variable in the multiple regression model once the variability in fish mercury associated with pH was accounted for. Methylmercury added only an additional 2.7% to the variation explained by the model and was only a significant contributor to the model if DOC was also included. DOC was also a small but significant component of the model.

Given the strong influence of lake water chemistry on methylmercury concentrations in both lake water and fish, it is questionable whether a single water quality criterion for mercury in water is adequate to protect fish-eating wildlife. A single water quality criterion applied to lakes with varying water chemistry, food web structure, and watershed characteristics may fail to adequately protect wildlife in some situations while being overly protective in other cases. Given the complexity of the dynamics of the mercury cycle, clearly a more sophisticated approach is warranted.

CHAPTER 7.0

MODEL VALIDATION

A validation step was initiated that examined the consistency and potential for generalization of the statistical relationships observed between fish mercury and water chemistry variables for the lakes sampled during the field study (Chapter 6.0). To achieve this validation, an independent dataset was obtained from the U.S. Geological Survey with a similar group of variables as collected during the field study. The data were collected from 106 sites from 21 National Water Quality Assessment (NAWQA) basins for the National Mercury Pilot Study. The purpose of the study was to examine relationships between total mercury and methylmercury in water, sediment, and fish. Parameters collected in surface water include total and methylmercury levels, total organic carbon, pH, and sulfate (SO_4). In sediment, MeHg and total mercury, and % loss on ignition (LOI) was collected. In fish, the weight, length, species, age, and mercury content of fish filet data were collected. These data are described in greater detail in Brumbaugh *et al.* (2001) and Krabbenhoft *et al.* (1999). The purpose of this section is to describe the fish[Hg]-water chemistry relationships observed with the independent USGS data and investigate their consistency, or lack thereof, with the data collected from the field study lakes.

This validation was approached in three general ways:

1. The same water chemistry variables that most significantly regressed with standard-length fish Hg concentrations in the field study lakes should reflect the same general influence (regression coefficients being either positive or negative) on fish Hg measurements from the validation lakes.
2. If the water chemistry data and models from the field study lakes are used to predict the MeHg concentration in the validation lakes, the predicted MeHg concentrations should approximately equal those concentrations of MeHg measured in the validation lakes.
3. If the regression models of fish Hg on water chemistry variables from the field study lakes are used to predict the average fish Hg concentration in the validation lakes, the predicted fish Hg concentrations should approximately equal those concentrations of Hg measured in fish from the validation lakes.

7.1 Methods

7.1.1 General

The data, as received from the USGS, are included in Table 7-1. To reduce the statistical influence of having inter-species differences (habitat, ecology, trophic status, etc.) between the datasets, only those lakes from which largemouth bass and smallmouth bass were collected were considered for further analysis. These two species were collected from the largest number of lakes (n = 51).

Some of the lakes represented in the USGS dataset were present up to three times in the dataset. This was a function of the manner in which data were reported by the USGS. To ensure that all lakes were represented equally within the dataset, that no bias existed with respect to certain lakes being represented more often than other lakes, the average fish length and fish Hg concentration were recalculated as the weighted average for any multiple listings. The revised dataset, used for the validation, is presented in Table 7-2. Only those water chemistry variables investigated in the field study lakes were included in the evaluation of USGS data. These included water concentration of methylmercury, water concentration of total mercury (TotHg), DOC, and pH.

Table 7-1. U.S. Geological Survey Data for Use in Validating the Fish Hg and Water Chemistry Variable Model.

Site Code	Site	Site Category	Species	No. Fish	Weight (g)	Length (mm)	Hgfish (µg/g ww)	Water MeHg (ng/L)	Water Total Hg (ng/L)	Water DOC (mg/L)	Water pH
TRIN	Trinity Crockett	Urb.	Largemouth Bass	1	311	260	0.127	0.015	6.5	5.8	7.5
TRIN	Trinity Crockett	Urb.	Flathead Catfish	3	191	262	0.110	0.015	6.5	5.8	7.5
TRIN	Livingston	Urb.	Largemouth Bass	1	1305	412	0.249	0.015	1.34	6	
TRIN	Livingston	Urb.	Largemouth Bass	1	849	366	0.373	0.015	1.34	6	
TRIN	Livingston	Urb.	Largemouth Bass	2	526	324	0.249	0.015	1.34	6	
TRIN	Livingston	Urb.	Largemouth Bass	2	306	278	0.125	0.015	1.34	6	
TRIN	White Rock Lk	Urb.	Largemouth Bass	2	341	282	0.051	0.015	1.24	5.6	7.9
TRIN	White Rock Lk	Urb.	Largemouth Bass	1	663	345	0.117	0.015	1.24	5.6	7.9
TRIN	Clear Crk	Bkg.	Spotted Bass	1	40	150	0.258				8
TRIN	Clear Crk	Bkg.	Spotted Bass	1	87	180	0.157				8
TRIN	Clear Crk	Bkg.	Spotted Bass	1	142	228	0.299				8
TRIN	Clear Crk	Bkg.	Channel Catfish	1	372	380	0.341				8
TRIN	Clear Crk	Bkg.	Channel Catfish	2	114	248	0.133				8
ALMN	Youghiogheny	Mine	Smallmouth Bass	1	292	290	0.072	0.016	1.16	2.05	7.6
ALMN	Allegheny	Urb.	Smallmouth Bass	3	108	206	0.087	0.006	0.85	1.82	7.8
ALMN	Dunkard	Mine	Smallmouth Bass	4	130	223	0.204	0.037	0.64	2.56	7.5
ALMN	Tenmile	Bkg.	Smallmouth Bass	2	430	312	0.288	0.104	2.48	3.87	7.7
ALMN	Tenmile	Bkg.	Largemouth Bass	1	592	329	0.181	0.104	2.48	3.87	7.7
ALMN	Clarion	Mine	Smallmouth Bass	2	745	363	0.327	0.093	8.09	4.17	7.4
LINJ	Passaic	Urb.	Pickerel	2	81	229	0.340	0.244	2.72	5.5	6.9
LINJ	Passaic	Urb.	Largemouth Bass	2	179	212	0.262	0.244	2.72	5.5	6.9
LINJ	Swan	Urb.	Pickerel	7	36	166	0.073	0.059	2.13	2.4	6
LINJ	Muddy	Ag.	Pickerel	2	307	345	0.159	0.061	2.25	3.9	7.3
LINJ	Muddy	Ag.	Largemouth Bass	2	362	295	0.334	0.061	2.25	3.9	7.3
LINJ	Muddy	Ag.	Largemouth Bass	2	110	192	0.207	0.061	2.25	3.9	7.3
LINJ	Great Egg	A/F	Pickerel	2	172	284	0.816	0.345	12.26	3.4	5.8
LINJ	Great Egg	A/F	Pickerel	5	84	237	0.495	0.345	12.26	3.4	5.8
LINJ	Great Egg	A/F	Largemouth Bass	1	49	148	0.575	0.345	12.26	3.4	5.8
SACR	Bear	Mine	Largemouth Bass	1	518	325	1.119	0.238	17.82	3.5	7.43
SACR	Bear	Mine	Smallmouth Bass	1	467	288	1.050	0.238	17.82	3.5	7.43

Table 7-1. U.S. Geological Survey Data for Use in Validating the Fish Hg and Water Chemistry Variable Model (continued).

Site Code	Site	Site Category	Species	No. Fish	Weight (g)	Length (mm)	Hgfish (µg/g ww)	Water MeHg (ng/L)	Water Total Hg (ng/L)	Water DOC (mg/L)	Water pH
SACR	Bear	Mine	Smallmouth Bass	1	150	220	0.499	0.238	17.82	3.5	7.43
SACR	Bear	Mine	Mountain Whitefish	1	549	390	0.246	0.238	17.82	3.5	7.43
SACR	Putah	Mine	Largemouth Bass	5	62	158	0.247	0.054	4.1	2.3	8
SACR	Cottonwood	Mine	Smallmouth Bass	1	401	315	0.264	0.03	1.02	1.5	8.5
SACR	Cottonwood	Mine	Smallmouth Bass	5	115	193	0.092	0.03	1.02	1.5	8.5
SACR	Sacramento	Ag.	Largemouth Bass	1	1471	550	1.803	0.148	10.19	3.2	7.9
SACR	Sacramento	Ag.	Largemouth Bass	1	1156	390	0.633	0.148	10.19	3.2	7.9
SACR	Sacramento	Ag.	Smallmouth Bass	1	675	324	0.466	0.148	10.19	3.2	7.9
SACR	Colusa	Ag.	Largemouth Bass	4	644	343	0.393	0.084	6.9	5.9	7.9
SANT	Fairview	A/F	Largemouth Bass	1	907	410	1.803	0.315	3.767592879	4.8	6.1
SANT	Branchville	A/F	Largemouth Bass	1	84	180	0.626	1.481	9.410022483	3.8	6.4
SANT	Springfield	A/F	Largemouth Bass	1	517	310	0.572	0.41		3	6.3
SANT	Canaan	A/F	Largemouth Bass	1	83	170	0.545	0.399	7.27	5.9	6.4
SANT	Saluda	Bkg.	Largemouth Bass	1	900	390	0.471	0.089	1.05	2.9	7.1
ACAD	Mermentau	Urb.	White Crappie	8	169	232	0.299	0.091	1.11	8.9	7.46
ACAD	Calcasieu	A/F	Sunfish	8	75	155	0.431	0.151	3.68	5.2	6.8
ACAD	Turtle	Bkg.	Largemouth Bass	8	387	292	0.200	0.031	1.03	8.6	7.4
ACAD	Lacassine	Ag.	Sunfish	8	132	173	0.645	0.458	1.61	11.1	6.66
ACAD	Tangipahoa	?	Largemouth Bass	8	562	335	0.743				
ACAD	Teche	?	Largemouth Bass	8	334	298	0.260				
ACAD	Falaya	?	Largemouth Bass	8	392	306	0.791				
ACAD	Segnette	?	Largemouth Bass	8	487	326	0.208				
CKIB	Chester	Urb.	Slimy Sculpin	5	20	40	0.036	0.019	2.96	3.8	8
CKIB	Campbell	Bkg.	Slimy Sculpin	5	23	45	0.069	0.019	2.5	1.6	7.8
CKIB	Campbell	Bkg.	Dolly Varden	5	141	233	0.086	0.019	2.5	1.6	7.8
CKIB	Deshka	Bkg.	Slimy Sculpin	10	35	60	0.060			8.4	6.8
CKIB	Costello	Bkg.	Dolly Varden	1	35	158	0.105	0.017	4.97	0.7	8.1
CKIB	Costello	Bkg.	Dolly Varden	4	11	114	0.064	0.017	4.97	0.7	8.1
NECB	Neponset	Urb.	Sunfish	4	50	127	0.309	0.278	4.4	6.6	6.6
NECB	Ipswich	Urb.	Sunfish	5	33	114	0.849	0.438	2.72	7	6.6

Table 7-1. U.S. Geological Survey Data for Use in Validating the Fish Hg and Water Chemistry Variable Model (continued).

Site Code	Site	Site Category	Species	No. Fish	Weight (g)	Length (mm)	Hgfish (µg/g ww)	Water MeHg (ng/L)	Water Total Hg (ng/L)	Water DOC (mg/L)	Water pH
NECB	Saugus	Urb.	Sunfish	4	19	100	0.287	0.11	2.79	4.4	7
NECB	Aberjona	Urb.	Sunfish	5	23	94	0.055	0.075	9.11	4	6.5
DELR	Neshaminy	Ag.	Smallmouth Bass	2	180	236	0.268	0.103	4.08	4.88	7.8
DELR	Tulpehocken	Ag.	Smallmouth Bass	1	354	283	0.236	0.089	2.14	2.69	7.9
DELR	Hay	Ag.	Smallmouth Bass	1	557	335	0.333	0.038	0.77	1.35	7.8
DELR	Hay	Ag.	Smallmouth Bass	2	221	258	0.290	0.038	0.77	1.35	7.8
DELR	Manataway	Urb.	Smallmouth Bass	3	206	253	0.169	0.061	1.37	3.32	8.7
DELR	Raccoon	Ag	Pickerel	1	330	372	0.288	0.049	1.11	2.86	7.3
GRSL	Cub	Ag.	Largemouth Bass	2	136	204	0.271	0.028	2.6	2.1	8.5
GRSL	Weber	Mine	Mountain Whitefish	1	833	391	0.141	0.095	21.76	3.1	8.5
GRSL	Weber	Mine	Mountain Whitefish	2	545	374	0.117	0.095	21.76	3.1	8.5
GRSL	Weber	Mine	Mountain Whitefish	2	317	300	0.073	0.095	21.76	3.1	8.5
LTEN	Seqwatchie	A/F	Largemouth Bass	1	228	250	0.421	0.014	1.38	1.37	7.6
LTEN	Seqwatchie	A/F	Largemouth Bass	1	114	214	0.262	0.014	1.38	1.37	7.6
LTEN	Buffalo	Bkg.	Largemouth Bass	1	351	282	0.241	0.014	1.45	0.85	7.2
LTEN	Buffalo	Bkg.	Largemouth Bass	1	132	212	0.390	0.014	1.45	0.85	7.2
LTEN	Buffalo	Bkg.	Largemouth Bass	1	35	144	0.217	0.014	1.45	0.85	7.2
LTEN	Indian	Urb.	Largemouth Bass	2	288	280	0.163	0.031	3.4	1.6	7.2
LTEN	Indian	Urb.	Largemouth Bass	1	470	324	0.256	0.031	3.4	1.6	7.2
LTEN	Indian	Urb.	Largemouth Bass	2	38	148	0.130	0.031	3.4	1.6	7.2
MIAM	Stillwater	Ag.	Smallmouth Bass	3	130	217	0.180	0.047	2.1	3.3	7.62
MIAM	Stillwater	Ag.	Smallmouth Bass	1	284	275	0.138	0.047	2.1	3.3	7.62
MIAM	TippCity	Ag.	Smallmouth Bass	4	178	242	0.117	0.05	2.52	3.4	7.94
MIAM	Mad	Ag.	Smallmouth Bass	1	518	325	0.271	0.084	0.79	2.1	7.96
MIAM	Hamilton	Urb.	Smallmouth Bass	1	345	292	0.113	0.194	3	5	8.4
MIAM	Whitewater	Ag.	Smallmouth Bass	1	80	187	0.135	0.025	0.81	1.9	7.7
MIAM	Whitewater	Ag.	Smallmouth Bass	1	129	220	0.128	0.025	0.81	1.9	7.7
MIAM	Whitewater	Ag.	Smallmouth Bass	2	257	270	0.198	0.025	0.81	1.9	7.7
MIAM	Milford	Urb.	Smallmouth Bass	3	55	153	0.042	0.049	1.57	3.5	8.3
MIAM	Milford	Urb.	Smallmouth Bass	1	698	340	0.276	0.049	1.57	3.5	8.3

Table 7-1. U.S. Geological Survey Data for Use in Validating the Fish Hg and Water Chemistry Variable Model (continued).

Site Code	Site	Site Category	Species	No. Fish	Weight (g)	Length (mm)	Hgfish (µg/g ww)	Water MeHg (ng/L)	Water Total Hg (ng/L)	Water DOC (mg/L)	Water pH
MIAM	Williamsburg	Ag.	Smallmouth Bass	1	107	205	0.205	0.066	2.09	4.4	8.3
MIAM	Williamsburg	Ag.	Smallmouth Bass	2	386	302	0.359	0.066	2.09	4.4	8.3
MIAM	Williamsburg	Ag.	Smallmouth Bass	1	608	350	0.520	0.066	2.09	4.4	8.3
MOBL	Shades	Urb.	Sunfish	8	29	118	0.122	0.035	1.31	2.56	8.2
MOBL	Satilpa	A/F	Largemouth Bass	1	92	194	0.551	0.072	2.09	3.84	7.5
MOBL	Satilpa	A/F	Spotted Bass	2	140	235	0.601	0.072	2.09	3.84	7.5
MOBL	Chickasaw	A/F	Sunfish	2	31	118	0.248	0.208	2.27	5.54	6.3
MOBL	Clairborne	Ag.	Crappie	2	210	245	0.146	0.04	1.778	4.4	7.8
MOBL	Clairborne	Ag.	Crappie	1	94	180	0.067	0.04	1.778	4.4	7.8
MOBL	Clairborne	Ag.	Spotted Bass	2	754	365	0.147	0.04	1.778	4.4	7.8
MOBL	Clairborne	Ag.	Spotted Bass	2	262	275	0.149	0.04	1.778	4.4	7.8
MOBL	Clairborne	Ag.	Spotted Bass	2	158	240	0.094	0.04	1.778	4.4	7.8
MOBL	Coosa	Ag.	Crappie	1	453	304	0.125	0.041	4.684	2.4	7.5
MOBL	Coosa	Ag.	Crappie	1	266	249	0.027	0.041	4.684	2.4	7.5
MOBL	Coosa	Ag.	Crappie	1	140	204	0.030	0.041	4.684	2.4	7.5
MOBL	Coffeeville	Urb.	Largemouth Bass	2	588	350	0.313	0.042	2.744	4.3	7.9
MOBL	Coffeeville	Urb.	Largemouth Bass	4	252	269	0.351	0.042	2.744	4.3	7.9
NROK	Turah Bridge	Mine	Mountain Whitefish	5	551	356	0.155	0.085	5.57	2	8.6
NROK	St. Regis	Ag.	Mountain Whitefish	5	264	294	0.092	0.021	1.53	1.5	8.1
OAHU	Wilson	Urb.	Tilapia	5	302	250	0.046	0.118	1.93	3.3	6.8
OAHU	Kawainui	Bkg.	Tilapia	5	127	188	0.030	0.018	1	3.1	6.95
OAHU	AlaWai	Urb.	Tilapia	5	124	179	0.020	0.014	1.17	0.6	8
OAHU	Nuuanu	Bkg.	Tilapia	5	249	226	0.063	0.101	24.27	2.5	7.83
OAHU	Hoomaluhia	Bkg.	Tilapia	5	49	133	0.018	0.013	0.6	1.2	7.68
SANA	Prado	Urb.	Sunfish	10	53	136	0.118	0.144	14.67	4.8	7.84
SANA	Chino-Corona	Urb.	Carp	5	393	311	0.114	0.082	2.99	6.9	10.09
SANA	Hamner	Urb.	Largemouth Bass	2	54	158	0.140	0.026	3.71	3	8.19
SANA	RefSite	Bkg.	Brown Trout	6	90	194	0.127	0.017	0.61	0.6	7.7
UIRB	Des Plaines	Ag.	Bowfin	1	617	337	0.174	0.101	4.18	11.4	7.49
UIRB	Nippersink	Ag.	Smallmouth Bass	2	84	180	0.066	0.041	1.42	3.53	7.81

Table 7-1. U.S. Geological Survey Data for Use in Validating the Fish Hg and Water Chemistry Variable Model (continued).

Site Code	Site	Site Category	Species	No. Fish	Weight (g)	Length (mm)	Hgfish (µg/g ww)	Water MeHg (ng/L)	Water Total Hg (ng/L)	Water DOC (mg/L)	Water pH
UIRB	Salt Creek	Urb.	Smallmouth Bass	3	70	175	0.088	0.132	9.26	6.27	7.29
UIRB	Pitner	Ag.	Largemouth Bass	3	154	219	0.129	0.026	0.27	3.1	7.78
UIRB	Mukwanago	Bkg.	Rock Bass	4	62	143	0.107	0.059	1.48	6.91	7.98
UIRB	Mukwanago	Bkg.	Largemouth Bass	1	114	182	0.075	0.059	1.48	6.91	7.98
YELL	Bighorn River	Ag.	Walleye	5	452	373	0.635	0.127	3.89	3.71	8.4
YELL	Bighorn Lake	Ag.	Walleye	5	896	453	0.676	0.095	2.48	3.77	8.3
YELL	Shoshone	Ag.	Walleye	1	1444	535	0.669	0.126	5.31	4.17	8.1
YELL	Shoshone	Ag.	Walleye	5	817	437	0.666	0.126	5.31	4.17	8.1
YELL	Tongue	Ag.	Smallmouth Bass	1	299	270	0.153	0.061	1.87	3.4	8.5
YELL	Yellowstone	Ag.	Sauger	2	176	294	0.250	0.153	4.07	2.9	8.6
SOFL	WC2-U3	Bkg*	Largemouth Bass	1	307	301	0.660	0.61	3.5	24	7.5
SOFL	WC2-U3	Bkg*	Largemouth Bass	1	227	273	0.370	0.61	3.5	24	7.5
SOFL	WC2-U3	Bkg*	Largemouth Bass	1	229	264	0.630	0.61	3.5	24	7.5
SOFL	WC3-3A15	Bkg*	Largemouth Bass	1	1694	507	4.220	0.5	1.9	16	7.1
SOFL	WC3-3A15	Bkg*	Largemouth Bass	1	205	250	0.820	0.5	1.9	16	7.1
SOFL	WC3-3A15	Bkg*	Largemouth Bass	1	466	325	1.420	0.5	1.9	16	7.1
NVBR	Lahontan	Mine	WhiteBass	8	694	366	3.360	0.92615	655.635	5.09	8.34

Table 7-2. Subset of U.S. Geological Survey Data to Validate Fish Mercury-Water Chemistry Relationships.

Site	Sample N	Fish Length Cm	Fish [Hg] ng/g	Water [MeHg] pg/L	Water [TotalHg] ng/L	Water [DOC] mg/L	pH
Allegheny	3	20.6	87.1	6	0.85	1.82	7.80
Bear	3	27.8	889.0	238	17.82	3.50	7.43
Branchville	1	18.0	626.0	1481	9.41	3.80	6.40
Buffalo	3	21.3	283.0	14	1.45	0.85	7.20
Canaan	1	17.0	545.0	399	7.27	5.90	6.40
Clarion	2	36.3	327.4	93	8.09	4.17	7.40
Coffeeville	6	29.6	338.0	42	2.74	4.30	7.90
Colusa	4	34.3	392.9	84	6.90	5.90	7.90
Cottonwood	6	21.3	121.0	30	1.02	1.50	8.50
Cub	2	20.4	271.0	28	2.60	2.10	8.50
Dunkard	4	22.3	203.9	37	0.64	2.56	7.50
Fairview	1	41.0	1803.0	315	3.77	4.80	6.10
Falaya	8	30.6	790.9				
GreatEgg	1	14.8	575.0	345	12.26	3.40	5.80
Hamilton	1	29.2	112.6	194	3.00	5.00	8.40
Hamner	2	15.8	139.8	26	3.71	3.00	8.19
Hay	3	28.4	304.0	38	0.77	1.35	7.80
Indian	5	23.6	168.0	31	3.40	1.60	7.20
Livingston	6	33	228.3	15	1.34		
Mad	1	32.5	270.6	84	0.79	2.10	7.96
Manataway	3	25.3	168.6	61	1.37	3.32	8.70
Milford	4	20.0	101.0	49	1.57	3.50	8.30
Muddy	4	24.4	271.0	61	2.25	3.90	7.30
Mukwanago	1	18.2	75.3	59	1.48	6.91	7.98
Neshaminy	2	23.6	268.4	103	4.08	4.88	7.80
Nippersink	2	18.0	66.3	41	1.42	3.53	7.81
Passaic	2	21.2	262.4	244	2.72	5.50	6.90
Pitner	3	21.9	129.4	26	0.27	3.10	7.78
Putah	5	15.8	246.8	54	4.10	2.30	8.00
Sacramento	3	42.1	967.0	148	10.19	3.20	7.90
Salt Creek	3	17.5	88.1	132	9.26	6.27	7.29
Saluda	1	39.0	471.0	89	1.05	2.90	7.10
Satilpa	1	19.4	550.9	72	2.09	3.84	7.50
Segnette	8	32.6	208.5				
Seqwatchie	2	23.2	342.0	14	1.38	1.37	7.60
Springfield	1	31.0	572.0	410		3.00	6.30
Stillwater	4	23.2	170.0	47	2.10	3.30	7.62
Tangipahoa	8	33.5	742.6				
Teche	8	29.8	260.0				
Tenmile	3	31.8	252.0	104	2.48	3.87	7.70
TippCity	4	24.2	116.9	50	2.52	3.40	7.94
Tongue	1	27.0	153.0	61	1.87	3.40	8.50
Trinity Crockett	1	26.0	126.8	15	6.50	5.80	7.50
Tulpehocken	1	28.3	236.0	89	2.14	2.69	7.90
Turtle	8	29.2	199.8	31	1.03	8.60	7.40
WC2-U3	3	27.9	553.0	610	3.50	24.00	7.50
WC3-3A15	3	36.1	2153.0	500	1.90	16.00	7.10
White Rock Lk	3	30.3	73.0	15	1.24	5.60	7.90
Whitewater	4	23.7	165.0	25	0.81	1.90	7.70
Williamsburg	4	29.0	361.0	66	2.09	4.40	8.30
Youghiogheny	1	29.0	72.4	16	1.16	2.05	7.60

7.1.2 Approach 1: Fish Hg versus Water Chemistry Variables

Fish length is the most statistically significant determinant of fish Hg concentration. In section 6.2, above, the statistical influence of fish length on mercury accumulation was controlled for by calculating the mercury concentration for a standard 10-cm length fish in each lake and then investigating the influence of water chemistry on the estimated mercury content of this standard length fish. Unfortunately, the data obtained from the USGS were not reported for individual fish specimens. Instead, the USGS reported the average length and average mercury concentration for multiple specimens collected from each lake. As a result, the Hg content of a standard-length fish could not be derived for each lake. Instead, the influence of fish length on fish mercury concentration had to be controlled across all lakes, rather than within individual lakes, prior to the investigation of the influence of water chemistry on fish Hg accumulation. When controlling fish Hg content for length across all lakes, any water chemistry or other variables that directly or coincidently correlate with fish length will have some of their influence on inter-lake variability in fish Hg levels explained or controlled. As a result, the subsequent regression analyses of length-controlled fish Hg levels on lake chemistry will have reduced capacity to explain inter-lake variability because some of it was already 'controlled' when controlling for fish length.

When $\log_{10}$(fish[Hg]) was regressed on log10(fish length), a quadratic regression model was found to provide the best overall representation of that statistical relationship (Figure 7-1).

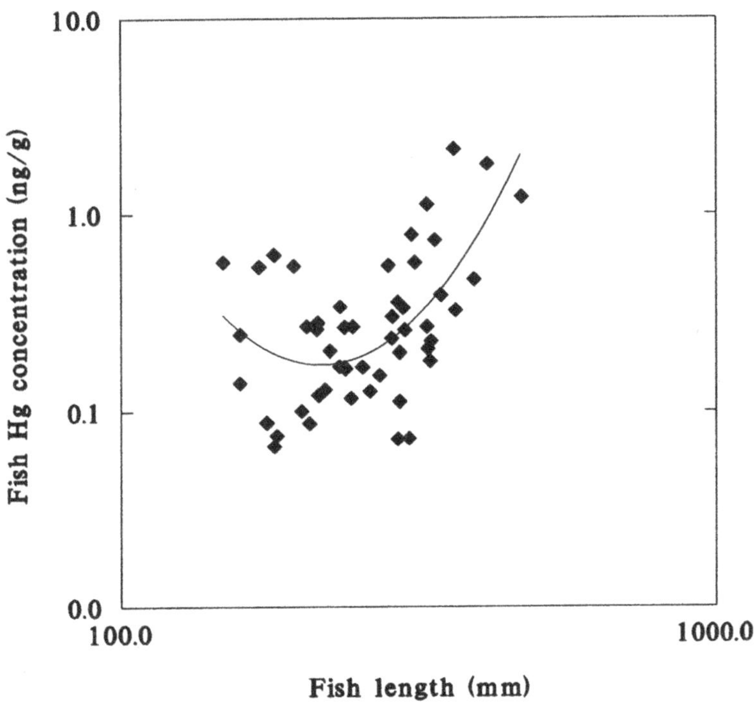

Figure 7-1. Regression of Average Fish Mercury Concentration on Average Fish Length for Largemouth Bass and Smallmouth Bass from 51 Lakes Surveyed by U.S. Geological Survey.

This was consistent with earlier investigations of the relationship of fish[Hg] on fish length across multiple lakes (Richardson, 1994). The residuals of that quadratic regression were then employed as the surrogate for the length-standardized fish[Hg] accumulation variable for subsequent regression on water chemistry parameters.

As mentioned above, this approach, dictated by the nature of the available USGS data, is valid but reduces the overall statistical strength of water chemistry-Hg accumulation relationships relative to the approach employed for the data from field study lakes, as reported in Chapter 6.0.

To investigate the influence of water chemistry on fish Hg accumulation, the fish[Hg]-length residuals were submitted to a step-wise multiple regression analysis on water chemistry variables. The water chemistry variables included pH, DOC, lake water MeHg concentration, and lake water total Hg concentration. All water chemistry variables but pH were $\log_{10}$-transformed to linearize their relationship with fish[Hg]-length residuals.

7.1.3 Approach 2: Predicting Lake Water MeHg in Validation Lakes from Study Lake Models

In the field study lakes, the best multiple regression model to predict lake water methylmercury concentration from other water chemistry variables was:

$$\begin{aligned} &Log(\text{MeHg}) = 1.324 + 0.456 \cdot Log\,DOC + 0.530 \cdot Log(\text{TotalHg}) \\ &R^2 = 0.318; p << 0.0005; n = 40 \end{aligned} \qquad (7.1.3\text{-}1)$$

where MeHg is the lake water concentration of MeHg (pg/L), DOC is lake water concentration of dissolved organic carbon (mg/L), and Total Hg is lake water concentration of total mercury (ng/L).

The regression model defined from the study lakes was then applied to estimate log(MeHg) values for validation lakes based on the DOC and Total Hg concentrations measured in those validation lakes. Subsequently, the estimated log(MeHg) values were regressed on the measured log(MeHg) values, fixing the constant at Y = X = 0, as the hypothesis being tested was that estimated log(MeHg) = measured log(MeHg), and the resulting slope of that regression was tested for its difference to a value of 1.00 (Y = X).

7.1.4 Approach 3: Predicting Fish Hg Concentrations in Validation Lakes from Study Lake Models

The testing of this hypothesis was more complicated. The validation lakes lacked sufficient data to permit derivation of the estimated mercury concentration in 10-cm standard length fish for direct comparison to the study lakes. Therefore, a different approach was necessary.

For each validation lake, the average length and average Hg concentration of all sampled fish (sample size per lake ranged from one to eight fish) were reported. The fish Hg and fish length data for the field study lakes were employed to estimate the $\log_{10}$ of the Hg concentration in fish with a $\log_{10}$ value of the standard length equal to the average fish lengths reported for each of 40 validation lakes (with all necessary water chemistry data). Then, still using the field study lake data, each of these estimated $\log_{10}$ fish Hg concentrations were regressed on pH, $\log_{10}$DOC, and $\log_{10}$MeHg, the three water chemistry variables that best predicted fish Hg concentration (see Chapter 6.0, above). The resulting regression coefficients were then employed to predict the $\log_{10}$ value of the average Hg concentration in fish for each validation lake (regression coefficients matched according to the average length of fish collected from each validation lake) based on the pH, DOC, and MeHg concentrations measured in those validation lakes.

Subsequently, the estimated logFishHg values predicted for the validation lakes were regressed on the measured logFishHg average values, fixing the constant at Y = X = 0, as the hypothesis being tested was that estimated log(FishHg) = measured log(FishHg), and the resulting slope of that regression was tested for its difference to a value of 1.00 (Y = X).

7.2 Validation Results

7.2.1 Approach 1

The relationship between average fish[Hg] and average fish length across 51 lakes sampled by USGS for largemouth bass and smallmouth bass, is presented in Figure 7-1. That relationship is best described by the following quadratic function:

$$\begin{aligned}&Log10(\text{fish[Hg]}) = 50.306 - 43.772 \cdot \text{Log}_{10}\text{Length} + 9.380 \cdot (\text{Log}_{10}\text{Length})^2\\&R^2 = 0.316,\ p << 0.0005, \text{n} = 51\end{aligned} \qquad (7.2.1\text{-}1)$$

The residuals from Equation 7.2.1-1, representing length-controlled fish Hg accumulation, were then submitted to both forward and backward stepwise multiple regression on water chemistry parameters. The resulting best model was:

$$\begin{aligned}&\text{Residual[Fish[Hg]} - \text{Length]} = 1.220 - 0.126 \cdot pH + 0.231 \cdot Log_{10}(\text{MeHg})\\&R^2 = 0.319,\ p << 0.005, n = 46\end{aligned} \qquad (7.2.1\text{-}2)$$

Notably, the lake water concentration of total Hg was not found to be a significant predictor of length-controlled fish[Hg], consistent with findings from the field study analysis. Length-controlled fish Hg accumulation was then regressed on the same combination of water quality variables found to be significant for the field study lakes. The results were:

$$\begin{aligned}&\text{Residual[Fish[Hg]} - \text{Length]} = 1.294 - 0.112 \cdot \text{pH} - 0.195 \cdot Log_{10}\text{DOC} + 0.296 \cdot Log_{10}\text{MeHg}\\&R^2 = 0.329,\ p << 0.005, n = 46\end{aligned} \qquad (7.2.1\text{-}3)$$

Although the addition of the DOC variable did not significantly improve the predictive power of the multiple regression, the nature (+/– direction) of the regression coefficients is the same as observed for the field study lakes. Both pH and DOC have an inverse (negative) influence on fish Hg accumulation, whereas lake water concentration of MeHg has a direct, positive influence.

7.2.2 Approach 2

When the estimated $\log_{10}$(MeHg) was regressed on measured $\log_{10}$(MeHg), fixing the Y intercept equal to zero, the resulting regression slope was 0.933 ($r = 0.96$; $p < 0.0005$; $n = 46$ lakes). This slope is not statistically different from 1.00 (t test; $t = 1.595$; $df = 44$; $p > 0.1$). These data are plotted in Figure 7-2.

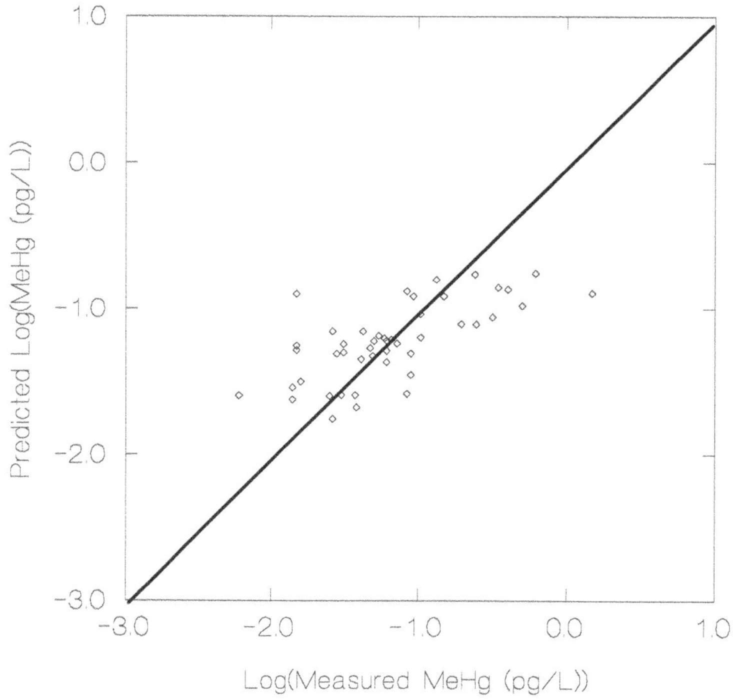

Figure 7-2. Regression of Estimated Lake Water Concentrations of MeHg on Measured Concentrations of MeHg for 46 Lakes Investigated by U.S. Geological Survey.
Note: The perfect fit line has been drawn with a slope of unity (Y = X).

7.2.3 Approach 3

When estimated $\log_{10}$(FishHg) was regressed on measured $\log_{10}$(FishHg), fixing the Y intercept equal to zero, the resulting regression slope was 1.117 ($r = 0.98$; $p < 0.0005$; $n = 44$ lakes). This slope is statistically different from 1.00 (t test; $t = 3.162$; $df = 42$; $p < 0.005$). These data are plotted in Figure 7-3.

Although the slope of predicted versus measured fish Hg for validation lakes was found to be statistically different from a value of 1.00, the measured slope is not numerically dissimilar for a slope of unity, representing an average error in prediction of only about 12%.

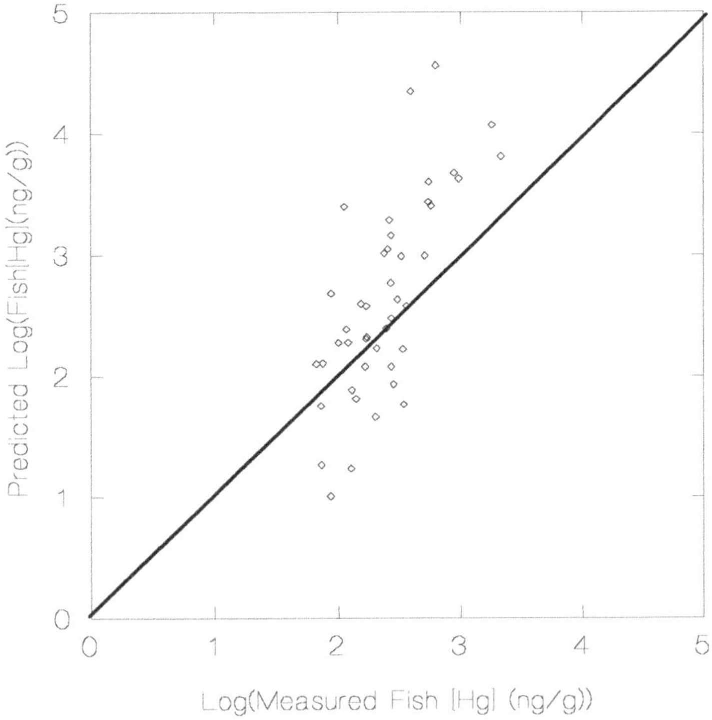

Figure 7-3. Regression of Estimated Average Concentrations of Hg in Fish on Measured Concentra-tions of Hg in Fish for 44 Lakes Investigated by U.S. Geological Survey.
Note: The perfect fit line has been drawn with a slope of unity (Y = X).

7.3 Discussion

The influence of water chemistry on fish Hg accumulation is not expected to be precisely identical from one lake, or one group of lakes, to another. Variations in physical, geological, and biological/ecological factors will prevent a precise replication of any particular regression relationship. Also, the necessary differences in the analysis and control of the fish[Hg]-length relationship between the USGS data and the field data preclude a precise replication of the subsequent analysis of water chemistry variables. However, it can be anticipated, and it is important, that the nature and direction of influence be the same.

From the data obtained from the USGS it was apparent that the influence of water chemistry variables on fish Hg accumulation follows the same general trends as those observed in the field study lakes. As pH declines, fish[Hg] increases; as DOC declines, fish[Hg] increases; and as lake water concentrations of methyl Hg increase, accumulation of Hg by fish also increases.

Also lake water concentrations of total Hg were found to have an insignificant influence on average, length-controlled fish Hg accumulation in both groups of lakes. This is consistent with this study's results (see Chapter 6), by Sorensen *et al.* (1990), and is further confirmation that the measurement and use of total Hg concentration in surface water, as an indicator of potential Hg accumulation risk to fish, piscivorous wildlife, and humans, is invalid and inappropriate.

The regression model developed from the field study data predicted fish Hg concentrations in the USGS lakes with reasonable accuracy. Although the models were found to slightly over-predict measured fish Hg concentrations, the error was minor, with the slope of the predicted-to-measured fish Hg concentrations being 1.117 (versus a desired slope of unity). It is likely that the predictive capability of the models would have been better if there had been more fish from each validation lake and if the individual length and mercury content of each fish had been reported.

These results suggest that the models developed from the field study lakes to estimate fish Hg concentrations may have wide application to predicting fish Hg levels in other lakes, as a function of lake water methyl Hg concentration, DOC, and lake water pH.

CHAPTER 8.0

Ambient Water Quality Criteria Model

8.1 Introduction

The U.S. EPA Ambient Water Quality Criteria for Mercury (U.S. EPA 1985) have not been updated in the years since being introduced in 1984 (2.4 μg/L as an acute [one-hour average] AWQC and 0.012 μg/L as the chronic four-day average). This may be primarily due to the subsequent development of AWQC for protecting wildlife (U.S. EPA 1997b; U.S. EPA 1995, 1996b) that are generally an order of magnitude more sensitive. Risk managers therefore are likely to use the AWQC for the protection of wildlife, as reliance on an AWQC for the protection of aquatic life would put wildlife species at risk. For this reason, an AWQC for the protection of aquatic life was not generated in this report.

The recent U.S. EPA Mercury Report to Congress (U.S. EPA 1997a) proposed a wildlife AWQC for mercury of 0.64 ng/L total mercury or 0.05 ng/L methylmercury. This wildlife criterion has a mix of assumptions that range in their degree of conservatism and toxicological uncertainty factors of unknown scientific rationale or unquantifiable uncertainty. The net impact of this mix of assumptions and uncertainty factors is that the level of protection afforded by the derived mercury AWQC to wildlife is unknown.

A better approach would be to develop a model that reflects lake-to-lake variation of factors particularly influential to mercury biogeochemistry and fish accumulation. Such a model would permit the derivation of lake-specific water quality criteria for mercury that are protective of piscivorous wildlife, while acknowledging the variability of the numerous physical, chemical, and ecological factors among lakes that ultimately influence mercury loading to lakes, methylation, bioavailability, bioaccumulation in fish, and wildlife exposure. Also, the uncertainties inherent in such a model (including measured variability in receptor characteristics of subject wildlife species, reported variability in toxicological response to mercury exposure, variability in lake morphometry, and water chemistry) will permit the quantification of the uncertainties present in both the developed model and in the national water quality criterion for mercury derived by the U.S. EPA (1997a).

Methylmercury concentrations in fish consumed by piscivorous wildlife are influenced by a variety of factors including lake water levels of DOC and sulfate, lake pH and alkalinity, food chain length, as well as a number of physical attributes of lakes that influence mercury influx (Brumbaugh *et al.* 2001; Richardson 1994; Richardson and Currie 1995; Richardson, Egyed, and Currie 1995a; Cabana *et al.* 1994).

Although the concentration of methylmercury in lake water is ultimately at the root, methylmercury accumulation in fish and subsequent wildlife exposure through fish consumption, the water concentration of methylmercury, or even total mercury has been the subject of scant research in relation to fish mercury accumulation. This has resulted, in part, from the relatively low concentrations of total mercury and methylmercury in surface waters, levels generally below routine analytical methods and detection limits until recently. This is coupled with the complex interactions of methylation/demethylation rate (whether in the water column or sediments), with water and sediment chemistry, sediment redox potential, and other factors (Hurley *et al.* 1998; Gilmour *et al.* 1998; Gilmour and Henry 1991; Beijer and Jernelov 1979; Miskimmin, Rudd, and Kelly 1992; Weber 1993; Winfrey and Rudd 1990; Xun, Campbell, and Rudd 1987).

Sorenson *et al.* (1990) published a study in which water levels of mercury (total) were collected in addition to the numerous water chemistry, food chain, and lake physical attributes that have been repeatedly shown to influence mercury accumulation in fish. In that study, data on 40 variables collected from 80 lakes in northeastern Minnesota were subjected to biostatistical analysis to ascertain their relative statistical influence on mercury concentrations in northern pike (*Esox lucius*) of 55 cm standardized length. The study found that lake water Hg concentration was an insignificant predictor of fish mercury concentration when combined in multiple regression with DOC and lake pH. Also, simple regression analyses demonstrated that DOC, pH, and zooplankton Hg concentration had a greater influence on fish mercury concentration than did water concentration of total Hg.

The significance of DOC to fish mercury accumulation has been noted by several authors (Wren *et al.* 1991; McMurtry *et al.* 1989). This influence likely relates to the role of DOC in the transport of mercury from watersheds into lakes via precipitation runoff (Lee and Iverfeldt 1991; Mierle and Ingram 1991; Richardson and Currie 1995), its role in complexation and bioavailability of Hg in the water column (Richardson and Currie 1995), and possibly abiotic methylation (Weber 1993). Consistently, DOC is one of the most significant determinants of Hg accumulation in aquatic biota (reviewed by Richardson 1994). The role of pH in fish mercury accumulation may relate to its influence on methylation rate (Gilmour and Henry 1991; Miskimmin, Rudd, and Kelly 1992; Xun, Campbell, and Rudd 1987) and also to its influence on DOC solubility, thereby indirectly affecting Hg bioavailability (Richardson and Currie 1995; reviewed by Richardson 1994). Food chain length has also been shown to be an important determinant of mercury accumulation in top predators of aquatic systems (Cabana *et al.* 1994).

The development of the mercury AWQC to protect wildlife consists of the following five steps:

Step 1: Derive a model that accounts for the site-specific water quality variables in relating levels of mercury in water to those in fish tissues.

Step 2: Using an independent dataset, evaluate and, if necessary, adjust the model from Step 1.

Step 3: Develop a dose-response model for mink exposed to mercury.

Step 4: Determine concentrations of mercury in water associated with a 10% probability of exceeding the LD_5 for mink for different site-specific conditions.

Step 5: Develop an equation that can be used to derive site-specific AWQC for mercury to protect wildlife.This chapter reviews each of the previously presented steps (Steps 1 and 2) and presents the methods and results of Steps 3 through 5.

8.2 Empirical Model Relating Mercury Levels in Water to Levels in Fish Tissue (Step 1)

Chapter 6.0 describes the development of a field study where numerous water quality variables were measured in 42 lakes in Ontario and Nova Scotia, Canada. These data were analyzed using a multiple regression analysis to determine the water quality variables that are the most important contributors to fish mercury body burdens. Based on the results of this analysis, an empirical model was developed that relates the mercury concentration in 10-cm fish from water chemistry variables (Equation 6.5.3-1)

8.3 Evaluation of the Empirical Model (Step 2)

Chapter 7.0 presents the results of the approach used to validate the empirical model using an independent data set provided by the U.S. Geological Survey (Brumbaugh *et al.* 2001; Krabbenhoft *et al.* 1999). It was apparent from the USGS that influence of water chemistry variables on fish Hg accumulation follows the same general trends observed in the field study lakes. As pH declines, fish mercury increases; as DOC declines, fish mercury increases; and as lake water concentrations of MeHg increase, accumulation of Hg by fish also increases. Lake water concentrations of total mercury were found to have an insignificant influence on average, length-controlled fish Hg accumulation. Finally, the empirical model developed in Chapter 6 (Equation 6.5.3-1) was able to predict the fish Hg concentrations in the USGS lakes with reasonable accuracy.

8.4 Dose-Response Model for Mink Exposed to Mercury (Step 3)

The dose-response model derived for mink exposed to mercury is an updated version of the model published previously by Moore *et al.* (1999a). The updated model described below takes advantage of a mink feeding study recently published by Dansereau *et al.* (1999).

For methylmercury, several long-term feeding studies (93 to 183 days) with female mink have been conducted using similar protocols (Aulerich, Ringer, and Iwamoto 1974; Wobeser, Neilsen, and Schiefer 1976ab; Wren *et al.* 1987ab; Chamberland, Belanger, and Dallaire 1996;

Dansereau *et al.* 1999). These studies were combined to produce a toxicity data set with 16 treatment levels. The endpoint was female mortality that, although not ideal, results in similar effects estimates as occur with reproductive endpoints such as kit biomass or number of surviving kits (Wren *et al.* 1987b). The concentration-response relationship for female mortality versus methylmercury concentration in the diet was estimated using the generalized linear model (GLiM) framework with both logit and probit link functions and a binomial error distribution. Kerr and Meador (1996) and Bailer and Oris (1997) describe this framework in detail. For these analyses, model fits were compared using log transformed concentration data and untransformed data. The analyses were carried out in SAS® and the 95% fiducial limits estimated using the following equations:

$$
\begin{aligned}
&FL = \frac{-b \pm \sqrt{b^2 - 4ac}}{2a} \\
&\qquad concn. = \frac{link\ p - \hat{\beta}_0}{\hat{\beta}_1} \\
&a = \hat{\beta}_1^2 - z2 \cdot (se\hat{\beta}_1)^2 \\
&b = 2 \cdot \hat{\beta}_0 \cdot \hat{\beta}_1 - 2 \cdot \hat{\beta}_1 \cdot link\ p - 2 \cdot z^2 \cdot se\hat{\beta}_0 \cdot se\hat{\beta}_1 \cdot corr(\hat{\beta}_0, \hat{\beta}_1) \\
&c = \hat{\beta}_0^2 + (link\ p)^2 - 2 \cdot \hat{\beta}_0 \cdot link\ p - z^2 \cdot (se\hat{\beta}_0)^2
\end{aligned}
\qquad (8.4\text{-}1)
$$

where $\hat{\beta}_0$ is the estimated parameter from the linear regression analysis, $\hat{\beta}_1$ is the estimated parameter from the linear regression analysis, *link p* is the inverse of the normal cumulative distribution for the probit link function or the logit tranformation for the logit link function (logit y – *p* / (1 – *p*), where *p* is proportion response), *se* is the standard error for the indicated parameters, *corr* is the correlation coefficient, and *z* is the normal deviate for α.

To convert effect concentrations to doses, the best-fit concentration-response model was combined with the food intake rate for captive female mink. Bleavins and Aulerich (1981) found that captive female mink had a mean food intake rate of 155.3 g/day (normalized to a 1 kg female mink). The coefficient of variation among individual mink for this study was small (6.4%), and thus this source of uncertainty was not considered further in the analysis.

The concentration-response model with the logit link function and log transformed concentrations data produced the best fit of the GLiM models that were considered. The log-logit model produced an adequate model fit to the data ($G = 6.53$; $p = 0.925$) (Figure 8-1). The model parameters were $\beta_0 = 0.0211$, $\beta_1 = -10.3$, $se\beta_0 = 0.214$, $se\beta_1 = 1.94$, and $corr\beta_0\beta_1 = 0.341$.

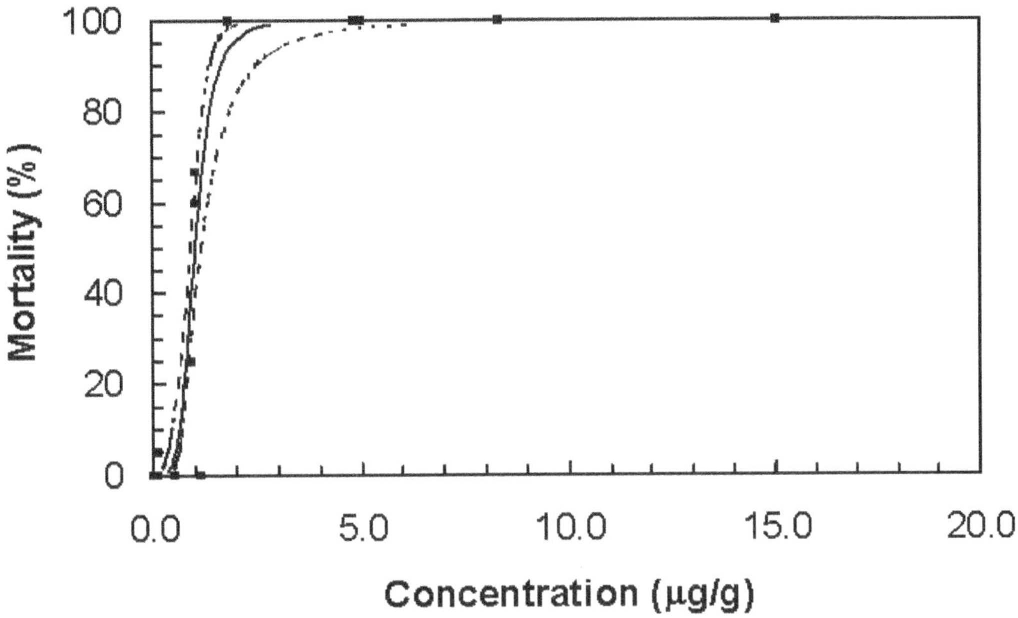

Figure 8-1. Concentration-Response Curve with 95% Fudicial Limits for Effects of Dietary Methylmercury on Survival of Female Mink During Chronic Exposures.
Note: Squares indicate original data.

To convert effect concentrations to doses, the log-logit concentration-response model was combined with the food intake rate for captive female mink. Bleavins and Aulerich (1981) found that captive female mink have a mean food intake rate of 155 g/day (normalized to a 1 kg female mink).The dose-response model is shown in Figure 8-2. The results indicate that the LD_5 for mink exposed to methylmercury in the diet is 80.5 µg/kg bw/day with 95% fiducial limits of 53.6 and 96.8 µg/kg bw/day.

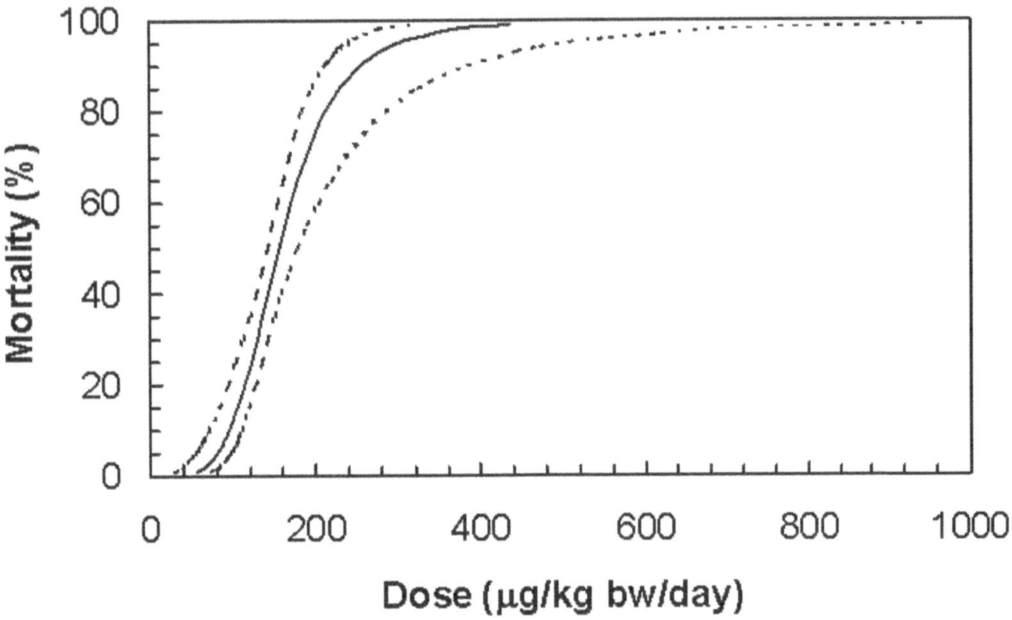

Figure 8-2. Estimated Dose-Response Curve with 95% Fiducial Limits for Effects of Dietary Methylmercury on Survival of Female Mink During Chronic Exposures.

8.5 Probabilistic Risk Model (Step 4)

A probabilistic risk model was used to estimate ambient water quality criteria for mercury for different combinations of the water quality parameters identified as having a significant influence on mercury levels in fish. Risk was estimated using the following equation:

$$RQ = \left(\frac{MR_{fw} \cdot C_f}{AE_f \cdot GE_f} \right) \div LD_5 \qquad (8.5\text{-}1)$$

where, RQ is the risk quotient (unitless), MR_{fw} is the metabolic rate of wild female mink (kcal/kg bw/day), C_f is the concentration of mercury in whole fish tissues normalized to 10 cm in length, AE_f is the food assimilation efficency for mink consuming fish (unitless), GE_f is the gross energy of fish, and LD_5 is the dose causing 5% mortality to female mink (mg/kg bw/day).

The exposure portion of the equation is typical of that used to estimate total daily intake for wildlife (U.S. EPA 1993c). This portion of the equation estimates mink exposure to methylmercury because this form of mercury is by far the most toxic (ATSDR 1989). This requires multiplying the variable for concentration of total mercury in fish (C_f) by the proportion of methylmercury in fish tissues (P_{MeHg}). The above equation also assumes that 100% of the mink diet is fish. This assumption is made for two reasons: 1) mink in some locations and at certain times of the year prey almost exclusively on fish (Alexander 1977) and AWQC should be protective of these situations, and 2) the wildlife AWQC is intended to be protective of other sensitive piscivores, such as river otter, that have a high proportion of fish in their diet (U.S. EPA 1993c).

To develop an AWQC for mercury, Equation 8.5-1 must be linked to mercury levels in water. This is done by substituting the empirical equation from the field study (Equation 6.5.3-1) to estimate levels in fish (C_f) from levels of mercury in water.

Input distributions were assigned as follows: lognormal distributions for variables that are right skewed with a lower bound of zero and no upper bound (metabolic rate), beta distributions for variables bounded by zero and one (assimilation efficiency of fish) or that are left skewed (proportion methylmercury in fish tissue, LD_5), and normal distributions for variables that are symmetric and not bounded by one (regression coefficients and constant in Equation 7.2.1-1, gross energy of fish). The distributions and distribution parameters used to generate the AWQC model are summarized in Table 8-1. The pH, DOC, and concentration of methylmercury in water (C_w) variables are user-defined deterministic values.

AWQC were derived for each of 77 different combinations of pH and DOC. The pH values considered ranged from 4 to 9 in increments of 0.5, and the DOC values considered ranged from 2 to 14 mg C/L in increments of 2. For each pH-DOC combination, different values were entered for C_w in an iterative series of analyses until the probability of the quotient exceeding 1 (which equals the probability that exposure would exceed the LD_5 for mink) was 10 ± 0.3%. The alternative approach of rearranging the risk quotient equation to isolate the C_w term on the left could not be used because Monte Carlo cannot be used to compute back calculations (Ferson 1996). For example, with the equation C (risk) = A (exposure) ÷ B (effects), one might solve for A by rearranging the equation to get $A = B \times C$ (similar to alternative approach for estimating C_w). In a deterministic analysis, this back calculation rearrangement works fine. In a Monte Carlo analysis, however, the answer is wrong, as can be seen by putting the distribution for B back into the original equation and computing C. The reason for the discrepancy is that A was computed by assuming independence between B and C. Risk (C), however, is not independent of effects (B). As Ferson (1996) states, "one is a function of the other, so independence is manifestly an inappropriate assumption."Solving such back calculation problems either requires the iterative approach used in this study or a special operation called deconvolution (Ferson 1995; Burmaster *et al.* 1995). The iterative approach used in this study involved assigning a value to C_w and conducting a Monte Carlo analysis with 5,000 trials and Latin Hypercube Sampling. If the probability of exposure exceeding the LD_5 was greater than 10.3% or less than 9.7%, lower or higher values of C_w, respectively, were used in subsequent iterations. To ensure that the final answer for C_w was stable, the final Monte Carlo analysis used 20,000 trials and Latin Hypercube Sampling.

Table 8-1. Input Variables Used in Monte Carlo Risk Quotient Simulations for Female Mink Exposed to Methylmercury.

Variable	Distribution	Parameters
Concentration of dissolved organic carbon in water, DOC (mg C/L)	Point estimate	User specified
pH in water, pH (unitless)	Point estimate	User specified
Concentration of methylmercury in water, C_w (pg/L)	Point estimate	User specified
Regression constant, C (unitless)	Normal	$\bar{x} = 3.033$, $s = 0.240$
DOC regression coefficient, RC_{DOC} (unitless)	Normal	$\bar{x} = 0.404$, $s = 0.170$
pH regression coefficient, RC_{pH} (unitless)	Normal	$\bar{x} = 0.161$, $s = 0.019$
C_w regression coefficient, RC_{Cw} (unitless)	Normal	$\bar{x} = 0.253$, $s = 0.111$
Metabolic rate of wild female mink, MR_{fw} (kcal/kg bw/day)	Lognormal	$\bar{x} = 256$, $s = 130$
Assimilation efficiency of fish, AE_f (unitless)	Beta	$\alpha = 65$, $\beta = 8$, scale = 1
Gross energy of fish, GEf (kcal/g)	Normal	$\bar{x} = 1.20$, $s = 0.240$
Proportion methylmercury in fish tissue, P_{MeHg} (unitless)	Beta	$\alpha = 38.2$, $\beta = 1.25$, scale = 1
5% lethal dose to mink, LD_5 (mg/kg bw/day)	Beta	$\alpha = 14$, $\beta = 5$, scale = 111

Figure 8-3 shows the results of a typical analysis, in this case for a pH of 5, and DOC of 12 mg C/L.

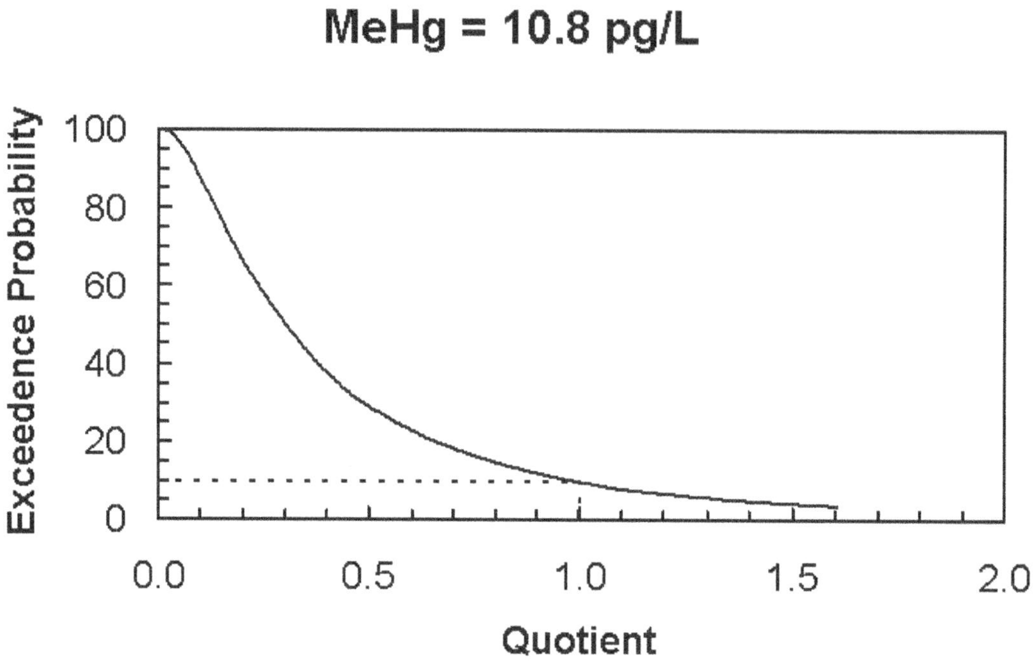

Figure 8-3. Risk Quotient Distribution for pH 5 and DOC Concentration of 12 mg C/L.
Note: The concentration of methylmercury in water that produced a 10% probability of the risk quotient exceeding 1 for wild female mink was 10.8 pg/L.

In this example, the concentration of methylmercury in water that produced a 10% probability of the quotient exceeding 1 (exposure exceeding the LD_5) was 11 pg/L. The results of the sensitivity analysis for this particular example indicated that seven input variables had rank correlations exceeding ±0.1: regression model constant ($C = 0.56$), metabolic rate for wild female mink ($MR_{fw} = 0.48$), regression coefficient for DOC ($RC_{DOC} = -0.43$), regression coefficient for concentration of methylmercury in water ($RC_{Cw} = 0.26$), regression coefficient for pH ($RC_{pH} = -0.21$), gross energy of fish ($GE_f = -0.20$), and 5% lethal dose ($LD_5 = -0.13$). The results of the sensitivity analysis varied, however, depending on the pH-DOC combination used (Table 8-2). For the regression coefficients for DOC, pH, and MeHg in water, the magnitude of the rank correlations increased as the user-specified values for DOC, pH, and MeHg in water increased, respectively (Table 8-2). The importance of other input variables (gross energy of fish, LD_5, metabolic rate of wild female mink) was less affected by changes in the user-specified parameters.

The remainder of this section includes additional notes on each of the input variables used in the risk quotient simulations.

Concentration of methylmercury in water (C_w), dissolved organic carbon, and pH. These variables are user-entered deterministic values.

Table 8-2. Results of Selected Sensitivity Analyses for Different User-Specified Combinations of pH, DOC Concentration in Water (mg C/L), and MeHg Concentration in Water (pg/L).

Variable	User-Specified Parameters			
	pH = 4, DOC = 2, MeHg = 0.29	pH = 9, DOC = 2, MeHg = 140	pH = 4, DOC = 14, MeHg = 3.6	pH = 9, DOC = 14, MeHg = 1100
C	0.66	0.52	0.59	0.44
MR_{fw}	0.56	0.45	0.5	0.37
RC_{DOC}	–0.14	–0.11	–0.46	-0.35
RC_{Cw}	–0.16	0.49	0.14	0.59
RC_{pH}	–0.2	–0.38	–0.19	–0.31
GE_f	–0.24	–0.21	–0.21	–0.16
LD_5	–0.16	–0.11	–0.13	–0.12

Regression constant (C) and coefficients (R_{pH}, R_{DOC}, RC_{Cw}). The distributions for these variables were based on the results of the multiple regression analysis described in Section 6.5.3. In each case, the summary statistics for each of these variables (the mean and standard deviation) were used to parameterize a normal distribution. The values used are listed in Table 8-2.

Assimilation efficiency and gross energy of fish (AE_f, GE_f). Metabolizable energy of a prey item is calculated by multiplying the assimilation efficiency for a predator consuming the item by the gross energy of the item. The food intake rate for each prey item is then estimated by dividing the metabolic rate (MR_{fw}) by metabolizable energy. The assimilation efficiency of fish consumed by mink was estimated by Moore, Breton, and Lloyd (1997) and Moore *et al.* (1999a), based on best professional judgment after examination of the relevant information in the Wildlife Exposure Factors Handbook (U.S. EPA 1993c). The mean assimilation efficiency of fish consumed by mink was estimated to be 0.91 with a standard deviation of 0.04. The gross energy of fish has been measured in numerous studies and has a mean of 1.2 kcal/g ww and a standard deviation of 0.24 (U.S. EPA 1993c).

Metabolic Rate (MR_{fw}). The mean metabolic rate of wild female mink standardized to 1 kg body weight can be estimated using the appropriate allometric equations developed by Nagy (1987). The 95% confidence limits are based upon the observed body weights in the field study by Mitchell (1961) and the following equation outlined in U.S. EPA (1993c) (Equation 8.5-2):

$$95\%\,CI = \log MRfw \pm c\sqrt{d + e(\log Wt - (\log Wt)^2} \tag{8.5-2}$$

where, log *Wt* is the mean weight of non-herbivirous mammals used in deriving the allometric equation and *c*, *d*, *e* are values of 0.321, 1.035, and 0.027 for mink (U.S. EPA 1993).

Proportion methylmercury in fish. Bloom (1992) found that close to 100% of the mercury in fish was methylmercury, which was consistent with the study results of Grieb *et al.* (1990) who found that 99% of the mercury found in samples of yellow perch, northern pike, and white suckers was methylmercury. Similarly, Southworth *et al.* (1995) found that 96.5% of mercury in fish caught in East Fork Poplar Creek, Tennessee was present as methylmercury. The distribution parameters listed in Table 8-1 are based on the combined results from the above studies.

Lethal Dose to 5% of Mink (LD_5). The estimated LD_5 and the fiducial limits estimated using Equation 9 were used to parameterize a beta distribution. A beta distribution was used for this variable because the fiducial limits were left skewed relative to the LD_5. A beta distribution can take on a left skewed shape, while a two-parameter lognormal distribution (the only option available in Crystal Ball 2000) can only have a right-skewed shape.

8.6 Derivation of the Site-Specific AWQC Equation for Mercury to Protect Wildlife

For each pH-DOC combination considered, a concentration of methylmercury in water was determined that corresponded with a 10% probability of exposure exceeding the LD_5 for mink. Using the SAS PROC REG procedure (SAS Institute 1990), a multiple regression was performed to derive the AWQC equation for mercury that can be used for any user-specified combination of pH and DOC in water. The dependent variable in this analysis was the concentration of methylmercury in water associated with a 10% probability of exposure exceeding the mink LD_5. This level of risk is used to define the AWQC for mercury that will be protective of wildlife. The list of possible independent variables included: pH, DOC, pH^2, DOC^2, pH^3, DOC^3, $\log_{10}pH$, $\log_{10}DOC$, and pH × DOC. The input dataset included 77 sets of parameters (11 × 7 pH-DOC combinations). In PROC REG, the MAXR model selection option was used. This option uses forward selection to fit the best one-variable model, the best two-variable model, and so forth. At each level, input variables are switched in and out of the analysis until the combination that maximizes R^2 is determined.

The multiple regression analysis to determine the AWQC equation for methylmercury in water indicated that adding input variables to the best four-variable model produced no additional benefit in terms of improving R^2. The R^2 for the four-variable model was 0.9997. Using the best four-variable model, the AWQC equation for mercury for the protection of wildlife is:

$$Log_{10}\,AWQC_{MeHg} = -3.748 + 0.348pH + 2.491\,(Log_{10}\,pH) + 1.318\,(Log_{10}\,DOC) - 0.00204\,(pH \bullet DOC) \tag{8.6-1}$$

where $AWQC_{MeHg}$ is the ambient water quality criterion for methylmercury in water (pg/L) and DOC is the concentration of dissolved organic carbon in water (mg C/L).

Table 8-3 uses Equation 8.6-1 to derive AWQC for methylmercury for various combinations of pH and DOC.

Table 8-3. Ambient Water Quality Criteria for Methylmercury in Water (pg/L) to Protect Wildlife.

pH	Dissolved Organic Carbon (mg C/L)						
	2	4	6	8	10	12	14
4	0.333	0.8	1.32	1.85	2.39	2.93	3.46
4.5	0.664	1.59	2.6	3.64	4.68	5.7	6.7
5	1.28	3.05	4.97	6.92	8.87	10.8	12.6
5.5	2.41	5.72	9.27	12.9	16.4	19.8	23
6	4.45	10.5	16.9	23.4	29.7	35.7	41.3
6.5	8.07	18.9	30.4	41.8	52.8	63.2	72.8
7	14.4	33.7	53.8	73.6	92.5	110	126
7.5	25.4	59.1	94	128	160	190	217
8	44.4	103	163	220	274	324	368
8.5	76.6	177	278	375	465	546	617
9	131	301	472	633	781	913	1030

The steps undertaken in this study to revise the Great Lakes and U.S. EPA mercury AWQCs for the protection wildlife follow several of the recommendations made by participants at the SE-TAC Pellston Workshop entitled "Reevaluation of the State of the Science for Water Quality Criteria Development," June 25–30, 1998. The purpose of the workshop was to evaluate how ecological risk assessment concepts could be used to develop water quality criteria in a more integrated and efficient process. Specific recommendations that came out of this workshop of relevance to this paper include:

- For chemicals in which water quality variables have been shown to have consistent effects on bioavailability, bioaccumulation, and/or toxicity, water quality criteria should be expressed as functions that explicitly incorporate these variables.
- There should be less reliance on generic, order-of-magnitude uncertainty factors in deriving AWQC. Alternative approaches include the use of empirically based uncertainty factors, probabilistic methods to establish less generic or subjective uncertainty

factors from limited empirical data (Richardson and Burmaster 1996), and uncertainty analysis to estimate concentrations that provide a specified level of protection (in this case, 10% probability of 95% mink survival) for at-risk wildlife species.

- The results of other lines of evidence (mesocosm studies, field biomonitoring, field alidation studies of proposed AWQC) should be used as checks on the appropriateness of the proposed AWQC.

Each of these recommendations was incorporated in the development of a revised ambient water quality criterion for mercury to protect wildlife.

CHAPTER 9.0

Data Gaps and Recommendations for Continuing Research

Although the concentration of methylmercury in lake water is ultimately at the root of methylmercury accumulation in fish and subsequent wildlife exposure through fish consumption, the water concentration of methylmercury, or even total mercury, has been the subject of scant research in relation to fish mercury accumulation. This has resulted, in part, from the relatively low concentrations of total mercury and methylmercury in surface waters, levels generally below routine analytical methods and detection limits until recently. The results of the field study and the independent validation analysis indicate the importance of focusing on methylmercury levels in surface waters, not total mercury levels, when managing risks of mercury to wildlife. Similar conclusions have been reached by other researchers in recent studies (Krabbenhoft *et al.* 1999; Brumbaugh *et al.* 2001). Future research, including monitoring programs, should consider this when measuring environmental media with the intention of managing risk to aquatic life, wildlife, and humans.

Recent trends in development and application of AWQC for wildlife protection are based on total mercury values. This is partially because of the historic approach to mercury criteria development using total mercury (Great Lakes AWQC, USEPA AWQC) and, more practically, because of the costs associated with measuring methylmercury in environmental media and tissues. The authors strongly believe that the use of a methylmercury-based AWQC, particularly with a probabilistic component, is a substantial improvement on the total mercury approach and as such improves upon the scientific defensibility of U.S. EPA approaches currently being implemented (Great Lakes). The authors recognize that additional practical problems are created for some applications of the methylmercury AWQC (effluent limits). This is primarily because of the lack of appropriate analytical methods, for example, that are approved for determining discharger compliance with an effluent limitation for methylmercury. Further research may be required into methods for translating AWQC for methylmercury into effluent limits that are protective and with which discharger compliance is measurable.

Although it would appear, based on the data from Greenfield *et al.* (2001) and the USGS data from Brumbaugh *et al.* (2001) and Krabbenhoft *et al.* (1999), that the AWQC model has the potential to be generalized to a broad geographic area spanning lakes with a wide variability in water chemistry characteristics, those published data are inadequate to fully test and validate this study's model. Inadequacies in these data are introduced by the limited number of fish specimens collected from individual lakes (with as few as one fish specimen collected in some cases) and by the relatively large fish that were collected, average sizes that fall well beyond the preferred prey size of mink. As a result, efforts must be made to compile or collect fully comparable data for a worthwhile testing and validation of the proposed AWQC model described in this report. This data set would include appropriate water quality parameters (see Table 6-1) from a wider variety of water bodies (seepage lakes, impoundments, rivers, streams) and also examined individual whole body fish for common fish species of an appropriate size (5 cm to 20 cm) would be desirable to further validate the field study data set and subsequent series of equations for the generation of the AWQC.

The piscivorous mammal mink was used to generate an AWQC to protect wildlife because of the availablility of a large number of feeding studies and its well-known sensitivity to mercury. As such, the AWQC is assumed to be protective of other piscivorous wildlife species (river otter, osprey). Future research on the toxicity of mercury to other wildlife species would assist in the evaluation of this assumption.

A field study program that monitors for effects to aquatic life and wildlife after the application of the AWQC at several sites would be an appropriate way to measure the level of protection afforded by the AWQC. The monitoring program should be instituted at several sites with widely differing characetristics (high/low pH, high/low DOC) to examine the impact of changes in site-specific parameters on the AWQC equation.

APPENDIX A

ASSESSMENT OF EXISTING METHODS TO ANALYZE MERCURY FOR THE PROTECTION OF AQUATIC LIFE AND WILDLIFE

Sagar Thakali
Herbert E. Allen
Department of Civil and Environmental Engineering
University of Delaware

September, 2000

Introduction

Since the Minamata tragedy (D'itri 1972), considerable progress has been made in terms of the methods to measure and analyze the levels of mercury and methylmercury existent in a variety of environmental samples. Currently, the methods suitable only for scientific investigations dominate the literature, mainly due to the difficulty of determining ultra low levels of mercury. We have used the existing methods and techniques for measuring mercury species in various environmental samples and evaluated their efficacy for various purposes, namely, monitoring, setting effluent limits, and scientific studies.

Accurate data are required in any kind of measurement. If the concentrations are over estimated, the industry may have to incur unnecessary costs to curb the resulting high effluent concentration values, and the effluent concentration would be unnecessarily below the regulatory limit. In the case of underestimation, the effluent concentrations would exceed the regulatory limit, and the monitoring would fail to demonstrate the required actual concentrations. The effects of inaccurate data on scientific investigation need not be elaborated.

Accuracy of the data is especially important when dealing with a trace metal like mercury. Most of the difficulties in mercury and methylmercury analyses arise because mercury occurs in ultra trace amounts in natural environments. Contamination during collection and preparation of the sample and the limitations of the analytical techniques themselves are two major sources of errors in measuring the concentrations of mercury species in various environmental samples. Another confounding factor is the inter-conversion of mercury species during the analytical processes. The decomposition of methylmercury into elemental mercury and methylation of elemental mercury contribute toward an over or underestimation of the concentrations depending on which way the inter-conversion proceeds during the analyses.

Yet another problem that needs further attention is that of certified reference materials (CRMs). The existing CRMs for both methylmercury and total mercury are inadequate in checking the accuracy of the methods under investigation because: 1) they do not represent the wide spectrum of sample matrices, 2) their reported concentrations are based on a limited number of methodologies, and 3) most of the analyses are based on methods which have been found to be flawed.

Finally, as most of the detection methods are based on spectrometry of some sort, interference during detection and measurement may contribute toward inaccurate analyses. Efforts must be continued to mitigate the adverse effects of the aforementioned difficulties to achieve adequate analytical methods for mercury species. While the ultra clean conditions have been relatively easy to achieve, albeit impractical in routine applications, the prospect of better, mercury-specific analytical methods and the problem of inter-conversion are still being investigated. In fact, either getting rid of inter-conversion processes during the analyses or an accurate way to account for them has been the focus of much research in the recent years. A separate focus has also been on the analyses of methylmercury alone, as it is in this form that mercury is most harmful to human health. However, the analyses of total mercury also continue to receive attention for better understanding of its fate and conversion into its more hazardous compounds.

Background

Existing literature credits the Westoo Method (Westoo 1966, Westoo 1967) for initiating modern mercury analyses. In it, a sample of fish tissue is homogenized and treated with a hydrohalic acid (usually HCl) in conjunction with other reagents to free the mercurial forms from the ligands in the matrix. The mercury species thus freed are then partitioned into an organic solvent like benzene. The mercury species are then repartitioned into an aqueous phase by using a complexing reagent like cysteine. After acidification of the aqueous phase to liberate methylmercury, the organomercurial is finally re-extracted into a solvent like benzene. Then Gas Liquid Chromatography (GLC) is used to separate the methyl mercuric chloride (the final form of the methylmercury in the original sample), and an Electron Capture Detector (ECD) is used for detection and quantification.

While the treatments of the samples in the current methods to isolate mercury continue to follow in the footsteps of the Westoo Method, the detection systems are usually based on spectrometry. The credit for this advancement goes to the Hatch and Ott Method (Hatch and Ott 1968).

In this method, the sample is taken into solution by an oxidizing acid attack. Mercury is then reduced to elemental state and aerated from solution in a closed system. The mercury vapor passes through a quartz absorption cell of an atomic absorption spectrophotometer, where its concentration is measured.

The Hatch and Ott method is mercury specific and is free from interference due to organic matter or other volatile constituents. For these reasons, this method has some obvious advantages over the ECD detector originally used in the Westoo method. However, when samples constitute a complex matrix with large amounts of easily reducible elements, the Westoo treatment is required to reduce probable interfering agents. It is not surprising that almost all of the mainstream mercury analyses are based on some combination of these two methods.

Clean Conditions

The importance of quality data cannot be overemphasized in the current context of mercury analyses. How stringent should the clean conditions be? The answer depends on the quality of the data sought. Scientific studies investigating the fate of mercury in the environment will require data as accurate as possible to establish the correlation among different species, agents, environmental conditions, etc. Compliance studies would require data that are not gross over- or underestimations. In any case, clean conditions always augment the quality of the data. Clean conditions should be observed throughout the analytical processes, right from the start of sampling to the end of cleanup, to ensure a dependable and adequate result.

The importance of using clean techniques can be seen in the analysis of mercury in Vandercook Lake using progressively cleaner sample collection and handling techniques (Fitzgerald and Watras 1989). Fitzgerald and Watras reported the following Hg concentrations in pM: 1983 (~1200), 1985-86 (~300), and 1986 (2.5 ± 0.2). Furthermore, Driscoll *et al.* (1994) suggest using any mercury data prior to 1985 with caution. According to methods employing clean conditions, total Hg appears to be in the range 1 to 10 ng/L in most areas. Humic lakes waters may get up to 15 ng/L (Fitzgerald and Watras 1989).

Before discussing the sampling and sample handling, it would be appropriate to discuss the cleaning of all containers and accessories in the laboratory. Needless to say, contaminants will be introduced into the sample if the containers and the apparatus used in sampling are not pre-cleaned. All labware should be cleaned by a process similar to hot-oxidation technique (Bloom and Crecelius 1983). The bottles are filled with a mixture of the sample water, 2% HNO_3, and 2% hot oxidizing reagent (100 g each of $KMnO_4$ and $K_2S_2O_8$ in 2 L deionized water that has been purified with 5 mL each of 12% $NH_2OH•HCl$, 0.3 M $FeCl_3$ and 30% NaOH). Glass stoppers and small pieces of equipment are placed into a large beaker of oxidizing solution. They are then heated to 80°–90°C in a hot water bath and kept at this temperature for an hour. After the appearance of MnO_2 (3–5 hours depending on the temperature), the contents are all discarded and all the surfaces rinsed with 12% $NH_2OH•HCl$, followed by deionized water to remove MnO_2. The bottles are then placed into 6 N reagent grade HNO_3 for one week.

For general water samples, Bloom (1995) suggests cleaning Teflon, glass and quartz materials by heating for 48 hours to >70°C in concentrated HNO_3 (in Teflon vats for safety). If hot HNO_3 cannot be used, then 4 N HCl is a satisfactory alternative. After this cleaning, the sampling containers for use in Hg speciation should be filled with 0.5 percent HCl, and heated to 60°C overnight, rinsed again three times, and filled again with 0.5 percent HCl. This procedure ensures the elimination of residual oxidizers from the HNO_3 bath and prevents reduction of species to Hg(II) while in storage. All sampling tubing, preferably made of Teflon, should also be cleaned in a similar manner.

If the containers are to be used only for total Hg analyses, the later steps in the cleaning procedure can be relaxed a little. They can be rinsed three times with ultra-low Hg water, and filled with the same water with 0.5% HCl and stored. The lids should be tightened and opened when required using a stainless steel wrench specifically made for this job.

Sampling and Sample Handling

Even before the samples reach the laboratory for analysis, there are ample opportunities for both positive and negative contamination. The first factors that should be considered to minimize contamination and loss of the analyte are the sampling, the handling, and preservation of samples prior to analyses. Substantial studies have been done on sampling and preserving samples for mercury analyses. What follows will review the existing literature on the sampling and preservation of different environmental samples for mercury analyses.

Water Samples

Mercury concentrations in rivers, lakes, and oceans have consistently been found in the part per billion to sub-part per trillion range when clean methods have been used whereas effluent concentrations appear to range from 5 to 200 ng/L (Hunt 1998). Problems associated with the storage, sampling, and preservation of water samples will be discussed next.

Storage

Polyethylene bottles, which are durable and contain relatively low trace element impurity, have been found to be unsuitable for long-term storage of water samples for mercury analyses. Under similar conditions, Pyrex (borosilicate) glass performs better than polyethylene bottles for storage of water samples for mercury analyses. Bothner and Robertson (1975) compared polyethylene bottles and Pyrex glass bottles for their suitability as containers for storing solutions for mercury analyses. Seawater samples (30% salinity) were acidified to pH 1.5 with HCl, well mixed and stored in the aforementioned bottles. The water was later analyzed for mercury using flameless atomic absorption spectrometry.

Mercury (Hg^{2+}) concentration in the Bel-Art polyethylene bottle, on an open bench top, was found to have increased from 0.01 to 0.14 μg/L in 245 days while there was no change in the mercury concentration in the Pyrex glass in the same period of time. In a later experiment, samples of acidified, distilled water in Monsanto polyethylene bottles otherwise under identical conditions also increased in mercury concentration from 0.007 to 0.044 μg/L in 60 days. Some of the observed increase in the mercury concentration has been tentatively attributed to a slow leaching of mercury from the container walls by acid solution, as a sample digested for two hours in Monsanto polyethylene bottles, using hot oxidizing solution, increased in mercury concentration to levels higher than those observed in the long-term storage bottles.

In the same study, diffusion of mercury vapor from the ambient air through a polyethylene container, as another mode of contamination, was proven. A Bel-Art polyethylene bottle containing seawater acidified to pH 1.6, in a tightly closed 20-L bottle together with a vial containing a mercury droplet 15 cm away, showed a concentration increase of over 1000-fold (0.002 μg/L to 2 μg/L) after five days.

Ambe and Suwabe (1977) examined Pyrex glass ampoules, glass bottles with Teflon sheet covering the internal wall and sealed with polycarbonate caps, and high density polyethylene bottles for suitability for storing standard solutions of mercury. The concentrations of the mercury standards (500 and 1000 μg/L) with pH adjusted to 2-3, remained constant in Pyrex glass ampoules during the 13-month period, whereas in the polyethylene bottles, mercury loss was observed after one month.

Both studies indicate that Pyrex glass is superior to polyethylene bottles for storage of water samples for mercury analyses, especially for long-term storage. The seemingly contradictory decrease in the mercury concentration in the later study may possibly be due to the diffusion of mercury vapor through the container wall to the ambient air. This corroborates the previous claim that diffusion may contribute substantially to errors in mercury analyses if the samples have been stored in polyethylene bottles in a laboratory with mercury in the ambient air.

The decrease in mercury concentrations when spiked seawater is stored in polyethylene bottles could also be due to faults in the analytical procedure and not actual losses. In their study using a mercury radioactive isotope, Carr and Wilkniss (1973) showed that 80% of the total mercury spiked in the seawater was associated with particulates at the time of sampling. After eight days storage at pH 1 (HNO_3), only 10%–15% was associated with particulates. Based on this finding, they suggested that the method could not adequately determine mercury associated with particulates, because it is unable to reach the detection system under the conditions of the analysis.

In a study by Sanemasa *et al.* (1976) when spiked seawater solutions were not acidified or contained only 0.001 M HNO_3, there was essentially complete mercury loss within five days. However, quantitative recoveries were found when spiked samples were stored without acid for 5–10 days and then acidified to 0.1 M. Since the all measurements were done using CVAAS, the data seem to support that the loss of mercury from seawater during storage is due to its conversion to a state that cannot be measured by the AA method. Sanemasa *et al.* (1976) conjectured that the acidification of the sample may release the mercury, strongly adsorbed on the suspended colloids, solid particles, and container walls, into aqueous phase or may convert it to readily reducible states.

Since these studies bring into light the likelihood of particulate mercury not reaching the detector and the likely unsuitability of polyethylene containers for storage of natural water samples for mercury analyses, it is safer to use Pyrex glass containers. A study of mercury stability in distilled and Hamilton Harbor (Canada) water by Carron and Agemian (1977) also established the superiority of Pyrex glass. The effectiveness, in diminishing order, of container types they examined was: Pyrex = soft glass > Teflon = Polypropylene = linear polyethylene > common polyethylene. For these reasons, EPA Method 1631 also specifies Pyrex glass containers for water samples for mercury analyses.

In the light of all these studies, it is advisable to follow the recommendation of Stoeppler and Matthes (1978) and store water samples in brown Pyrex glass bottles with tight screw caps made of dense, rigid plastic material. The brown color is suggested to minimize the effect of light, if there is any, on the degradation of methylmercury. Furthermore, it always better if the analyses can be performed as soon as possible after sample collection; that is to say that storage time should be minimized, if it can not be totally prevented. Storage time should not exceed 10 days.

Sampling Devices

For sample collection, polyvinyl chloride Niskin type sampling bottles (Bothner and Robertson 1975) and Teflon-coated Go-Flo sampling bottles (Bloom and Crecelius 1983) have been used. The former were found neither to add mercury to unspiked distilled water nor to subtract mercury from water spiked to about 0.05 µg/L of Hg^{2+} in tests lasting 45 minutes (Bothner and Robertson 1975). The latter has been found to neither contaminate nor adsorb mercury from the seawater over the short term of a sampling period (~30 minutes) (Bloom and Crecelius 1983). As long as the samples are treated and transferred into sample bottles within about 30 minutes of collection, the type of sample device does not seem to cause a significant loss or contamination.

Colman and Breault (2000) used an all-Teflon isokinetic sampler cleaned at 75°C with 4 N HCl. Assessing the blank contributions, they found that the mean mercury concentrations in field-equipment blanks were 0.135 ng/L for total mercury and 0.0086 ng/L for monomethylmercury.

Preservatives

Aside from the storage material and the care in handling the samples, the type of reagents used for preservation, the acidity (pH) of the sample, the composition of the sample matrix, and physical conditions like temperature also affect the loss of mercury from the sample. Several studies have investigated the effect of pH on the preservation of mercury in water samples. Depending on the type of acid used, the composition of the sample matrix, and the concentration of the mercury in the sample, a low pH in the range of 0.5–2.0 seems to achieve the best results.

Ambe and Suwabe (1977) suggested pH adjustment to 0.5–1.0 using H_2SO_4. They found that at pH 1, a standard solution containing 20 µg/L mercury was stable for 21 months, whereas at pH 3, the concentration dropped by 80% in one month after preparation. However, a 1-µg/L standard solution at pH 1 showed a slight decrease after two months. Sanemasa *et al.* (1976) found that HNO_3 concentrations above 0.1 M effectively prevented loss of mercury spiked in seawater for a period of at least four months. HCl has also been commonly used for acidifying the samples (Bothner and Robertson 1975; Stoeppler and Matthes 1978). In fact, Stoeppler and Matthes (1978) found that when spiked seawater samples were acidified with concentrated HNO_3 or HCl to a pH of about 2.5, contrary to Ambe and Suwabe (1977), no change in mercury concentration within the analytical error of the AAS method was observed during a storage of about 2.5 months. However, methylmercury chloride was seen to degrade very rapidly under identical conditions when HNO3 was used for acidification. For this reason, HCl is deemed the better of the two for the long-term storage of seawater samples. It should be noted that acids are not used alone as preservatives. They are used in conjunction with other reagents, which will be discussed next.

Sodium chloride (NaCl) at concentrations above 0.01 M in cojunction with 0.1 M HNO_3 was found effective in preventing loss of mercury from spiked distilled water and natural waters stored in polyethylene bottles (Sanemasa *et al.* 1977). In a study Ambe and Suwabe (1977) found that when 1-µg/L solutions containing 3% NaCl adjusted to a pH below 1 with H_2SO_4 were sealed in glass ampoules, the concentration remained constant during 35 days of the test. Incontrast, the mercury concentration in solutions without NaCl, prepared as a reference, decreased rapidly. After similar tests, they showed that at least 2% NaCl is necessary to achieve a stable mercury standard. These studies support the assumption that formation of chloromercury complexes in acidified seawater and salt-amended distilled water augments the stability of mercury in them during storage (Sanemasa *et al.* 1976). However, as will be discussed later, the presence of chloride ions in spectrometric analyses introduces significant interference (Bloom 1989).

Lo and Wai (1975), in their storage tests of de-mineralized water spiked with ^{203}Hg at 5 µg/L and stored in Nalgene polyethylene bottles, found that HNO_3 at pH 0.5 and $K_2Cr_2O_7$

(0.05% by wt.) effectively inhibit adsorption of mercury to the container walls and volatilization, respectively. Thus, when used in conjunction, they are expected to prevent mercury loss due to the mechanisms mentioned above. Using the acid dichromate solution at pH 0.5, the loss of mercury from the water was reduced to only 2 ± 1% (as compared to 16% when only HNO_3 and 25% when only $K_2Cr_2O_7$ were used) after 21 days of storage. A solution of HNO_3 containing 0.2 µg/L Au^{3+} (in the form of $HAuCl_4 \cdot 3H_2O$) at pH 0.5 was as effective as the aforementioned acid dichromate solution.

Carron and Agemian (1977) examined six different preservatives: 5% HNO_3, 1% H_2SO_4, 5% HNO_3 + 0.005% $K_2Cr_2O_7$, 1% H_2SO_4 + 0.005% $K_2Cr_2O_7$, 1% H2SO4 + 5% HNO_3 + 0.005% $K_2Cr_2O_7$, and 2% HCl + 0.005% $K_2Cr_2O_7$) on synthetic water (deionized water) and natural water spiked with 0.20 and 0.60 µg Hg/L. Using the method of total errors (McFarren, Lishka, and Parker 1970) preservatives 3, 4 and 5 were found to be far more effective than the rest. As long as the pH was less than 1, the concentrations of acid in these three methods made little difference. They suggested the use of H_2SO_4 rather than HNO_3 based on its lower volatility, higher stability, and lower blank contribution; 1% H_2SO_4 + 0.005% $K_2Cr_2O_7$ was compatible with the automated Cold Vapor Atomic Absorption (CVAA) method with a blank contribution of less than 0.01 µg Hg/L. Furthermore, using 0.05% $K_2Cr_2O_7$ instead of 0.005% $K_2Cr_2O_7$ with 1% H_2SO_4, considerably improved the precision and accuracy.

The effectiveness of $K_2Cr_2O_7$ and Au^{3+} in preventing loss of mercury through volatilization could be due to their high oxidizing power which keeps mercury ions in the solution as Hg(II), thus preventing the disproportionation reaction of Hg(II) to Hg(I) and Hg(0) in the aqueous phase to form volatile mercury species (Lo and Wai 1975).

It seems that major loss of mercury species while storing water samples is due to volatilization and adsorption to the container walls. The adsorption to the container walls would be significantly reduced, as mentioned earlier, by using Pyrex glass as the storage container. To prevent further adsorption and volatilization, acid dichromate solution as the preservative is likely to achieve the best result. Organic preservatives like EDTA and cysteine (Weiss, Shipman, and Guttman 1976), which form extremely stable mercury complexes are not discussed as possible preservatives in this study because their use requires an additional step in the analyses to liberate the mercury species.

More recently, Bloom (1995) recommended storage conditions specified in Table A-1 for mercury species in water samples stored in Teflon bottles.

Table A-1. Recommended Storage Conditions and Storage Times for Different Mercury Species.

Sample Matrix	Storage Conditions	Storage Time
Total Hg	1% BrCl, $AuCl_3$, $K_2Cr_2O_7/H_2SO_4$	> 1 year
Methyl Hg	Freeze at –18°C*	> 1 year
Methyl Hg	0.5 – 1% HCl (dark)	6 months
Total and methyl Hg	No preservative, 0 – 4 °C, dark*	1 week**
Volatile species	No preservative, 0 – 4 °C, dark*	2 days**
Dissolved/particulate	No preservative, 0 – 4 °C, dark*	2 days**

Source: Bloom 1995.
* Just prior to analysis, 0.5% HCl is added and the sample is allowed to sit one day to release Hg from the walls of the vessel.
** Value is in use, but not rigorously verified.

Sediment Samples

For shallow waters, hand-held coring or grab-sampling devices have been used. In deeper waters, different core and grab samplers have been used. All of these sampling devices appear to be adequate for mercury analyses of the sediments (EPRI 1987). Care should be taken to obtain samples representative of a site. Concentrations across a single sediment bed may vary due to various eco-biological processes. Methylation of mercury has been found to occur in freshwater and marine sediments. This process is mediated both by bacteria (Jernelov, Ndner, and Rsson 1975) and by humic substances (Nagase *et al.* 1982). Several studies have reported 10- to 100-fold enrichments of mercury concentrations in pore waters extracted from marine and freshwater sediments (Bothner *et al.* 1980, Lu *et al.* 1986). The same studies suggest that these elevated levels of mercury in pore waters from anoxic sediments are likely to be present in the form of polysulfide complexes such as HgS_2^{2-} or HgS_2H^-. They suggest that very high total organic mercury concentrations (197 to 804 µg/kg) in sediments collected off the coast of southern California representing 32% to 59% of the total mercury reported by Eganhouse (1975) may be mercury bound to fulvic substances.

Given the complex nature of the mercury species and the associated processes, storing sediment samples without proper preservatives and techniques may result in measurements that are significantly different from their true values at the time of sampling.

Preservation and Storage of Sediment Samples

In order to prevent biological degradation by bacteria and inter-conversion of different mercury species, most studies measuring organic mercury freeze or ice-store sediment samples.

1. Bartlett, Craig, and Morton (1977) reported a two- to four-fold initial increase in methylmercury in stored, unfrozen sediment followed by a decrease in methylmercury concentration.
2. Eganhouse (1975) reported that repeated freezing and thawing of sediment may result in the loss of up to 96% of the organic mercury present in the sediments.

3. Horvat, Bloom, and Liang (1993) also investigated the behavior of total and methylmercury in sediment samples in different storage conditions. The concentrations are usually determined in a wet sample and then expressed on a dry weight basis after correction based on wet/dry weight ratio separately determined by drying the sediment to a constant weight at 85°C. Drying unspiked sediment samples at 85°C resulted in a 15% and 45% decrease of total and methylmercury, respectively. Similar results were obtained for sediment samples spiked with inorganic and methylmercury. Additionally, methyl and total mercury were found to be stable in fresh sediments for up to 1.5 months if refrigerated (0 to 2°C) and for up to six months if deep-frozen (–18 to –22°C). However, the authors recommended further investigations of these methods, including lyophilization.

Biological Samples

Biological samples constitute a wide variety of plant and animal tissues. Bioaccumulation of methylmercury in aquatic organisms is a well-known fact. Eighty-five percent of methylmercury in the aquatic system is found in fish (Hovart, Liang, and Bloom 1993). The main problem with sampling biological organisms arises due to the need for homogeneity. The concentration of mercury species varies not only across the organisms in different enclaves of the same aquatic system, but also across the different tissues in the organism itself. While a versatile procedure that is able to handle various tissues, bones, and fibers separately and/or together is desirable, such a procedure is bound to be too complex, too complicated, and probably too time consuming to be employed for routine applications.

Nichols and Hageman (1979) proposed a noncontaminating, representative sampling by shattering of cold, brittle, biological tissues. In this procedure, frozen samples weighing from 0.01 to 5 kg are placed in a plastic bag and immersed in liquid nitrogen until adequately chilled. The sample thus chilled is then placed into several layers of polyethylene bag. A layer of canvas is used to protect the outer layer of polyethylene bag from tearing. Using the blunt side of a hammer, the wrapped sample is then shattered into brittle fractures. After unwrapping the shattered sample, pieces are transferred into a tared flask with acid-cleaned plastic scoops and a glass rod.

A second procedure involving a heavy duty, high speed Waring blender was also used. In it a ~3 g sample is chilled with liquid nitrogen (or dry ice) and then placed in a 100-mL stainless steel blender container. The sample is blended into powder (10–30 seconds) and poured into a clean container where the dry ice is allowed to sublime.

More recently, Driscoll *et al.* (1994) handled fish samples in the following manner: fish were individually wrapped in aluminum foil, placed in Ziploc bags, and kept on ice until they could be frozen (within four to eight hours after collection).

Clean Laboratory Facilities

The main concerns in the laboratory while analyzing mercury species are the contamination through gaseous and particulate mercury species in the ambient air, the use of contaminated laboratory ware, and the use of impure reagents. Ferrari, Moreau, and Boutron (2000), in the context of ice and snow samples deal with these specific difficulties and propose a set of very stringent clean conditions for determination of ultra-low levels of mercury, which are equally applicable for the analyses of any trace element.

The diffusion of mercury in the ambient air into the sample may contribute significantly to the errors in the analyses. It has been reported that more than 95% of the total mercury in the atmosphere is present in gaseous forms (Slemr, Schuster, and Seiler 1985; Lindberg and Stratton 1998). The rest are in particulate forms. Several studies have reported the need for high efficiency particulate (HEPA) filters to filter the air used to flush the laboratory (Schroeder *et al.* 1995; Wang, Xiao, and Lindqvist 1995; Biester and Scholtz 1997).

The class 100 level is an accepted standard for environmental trace metal work. If a clean room is not available, a class 100 clean bench or nonmetal glove box, fed by mercury- and particle-free air or nitrogen, should be used for all sample preparation (EPA Method 1631). In the clean room, the tempered air from outside is filtered through high efficiency particulate (HEPA) filters into the room (Bloom 1995). Unlike other trace metals, more that 95% of mercury exists in gaseous phase. Thus it is a greater contaminating source than its particulate counterpart (Ferrari, Moreau, and Boutron 2000). Bloom (1995) offers the following guidance for clean rooms with gaseous mercury:

- If total gaseous mercury is under 10 ng/m^3, then the lab is sufficiently clean for mercury work.
- If it is between 10 and 100 ng/m^3, then it is provisional and extra precautions should be taken in sample handling to avoid contamination.
- If the lab air contains more than 100 ng/m^3 of gaseous mercury, then it should be considered unacceptable for environmental mercury analyses.

Obvious mercury sources like thermometers should be eliminated from the lab (Bloom 1995). Bloom also recommends an exchange rate for laboratory air of 10 room volumes per hour with the outside air using HEPA filters, assuming the outside concentration of mercury in the air is lower than that of the lab. Additionally, use of mercury-specific adsorbents like iodated carbon, sulfur-impregnated carbon or gold-impregnated materials is recommended. These are used together with HEPA filters and a pre-filter is placed in between.

In a similar manner, Ferrari, Moreau, and Boutron (2000) maintain that a clean laboratory should be positive pressure with non-laminar flow and have a Hg laminar flow clean bench. The clean benches should be made of polyethylene and washed with acidified pure water. Use of HEPA filters for particulate mercury elimination and charcoal filters to eliminate gaseous mercury is highly recommended.

Boutron (1990) describes the positive pressure non-laminar flow clean laboratory (Marsh 1980, Marsh 1981) and older clean benches. Their laboratory was enlarged and modified for more room in 1996–1997. The details including the area, size and constructing materials of the working surfaces and benches can be found in Boutron (1990) and Ferrari, Moreau, and Boutron (2000). Flushing the air through an activated charcoal filter, a prefilter (92% efficiency), and finally through a HEPA filter (99.999% efficiency), Ferrari, Moreau, and Boutron (2000) were able to achieve a concentration of 2.2 ± 0.6 pg/L (n = 18) for gaseous mercury in the laminar flow clean bench, as opposed to 12.4 ± 2.0 pg/L (n = 20) in the laminar clean room.

Last but not least, the clean lab should have provision for ultra clean water with which ultra pure reagents are prepared. The water is best produced using a series of activated charcoal and mixed bed ion exchange resin columns (Ferrari, Moreau, and Boutron 2000). Bloom (1995) also mentions the use of similar system for water production in a clean laboratory, starting with an industrial carbon bed to remove chlorine and mixed bed deionizing resins to remove trace ions. The water from this system is further passed to a smaller system like Milli-Q, NANO-Pure etc. The water thus produced should not have a mercury concentration of more than 1 ng/L.

Clean laboratories set up as just described are not always available and may even be too stringent and thus too costly to be employed for routine monitoring analyses. Nonetheless, the possibilities of contamination due to ambient air should not be ignored and should be minimized whenever possible.

Clean Garments and Gloves

In any case, clean conditions are imperative to determine ultra-low levels of mercury species. To achieve this effect the U.S. EPA Method 1631 (1996a) also requires similar conditions for analyses of mercury species in water. Use of clean, nontalc gloves at all times by the sampling personnel and change of gloves even in the case of suspected contamination is deemed imperative. Colman and Breault (2000) had all sampling and sample processing handled by personnel wearing Tyvek coveralls and shoulder-length plastic gloves using a clean protocol.

The use of clean gloves and garments in the clean lab has been deemed necessary by both Boutron (1990) and Bloom (1995). When entering a clean room, the personnel should change into clean room shoes and coveralls and walk over a "tacky mat" that removes dust from the shoes. Once in the lab, clean gloves should be used at all times and changed as necessary.

Detection Limits and Interferences

Basically, the analyses of mercury species involve three stages: sample treatment, separation from the matrix/isolation from other species, and detection. The methods for different environmental samples in general vary in the treatment and, to some extent, in the separation and isolation. Most of the detection systems are based on some variations of atomic spectrometry. For this reason, it might be useful to introduce Cold Vapor Atomic Absorption Spectrometry (CVAAS), Cold Vapor Atomic Fluorescent Spectrometry (CVAFS), and Inductively Coupled Plasma Mass Spectrometry (ICP-MS), which have been the methods of choice for regulatory purposes and scientific investigations, respectively.

Cold Vapor Atomic Absorption Spectrometry (CVAAS) (Skoog, Holler, and Neiman 1998)

Cold vapor atomization is specifically applicable only to mercury analyses because of its appreciable vapor pressure at ambient temperature. The cold vaporization is followed by atomic absorption spectrometry. The mercury in a sample is converted to Hg^{2+} by treating it with an oxidizing mixture of nitric and sulfuric acids. The resulting Hg^{2+} is then reduced to its elemental state (metal) with $SnCl_2$ (or other reducing agents). Then elemental mercury is swept into a long-pass absorption tube by bubbling a stream of inert gas through the mixture from which the elemental mercury was formed, and the absorption is measured at 253.7 nm.

Interference in AAS:

- *Spectral interferences:* The absorption or the emission of an interfering species either overlaps or lies so close to the analyte absorption or emission that resolution by the monochromator becomes impossible.
  Causes: Presence of combustion products that exhibit broadband absorption and particulate products that scatter radiation; matrix interference; carbonaceous particles are capable of scattering light.
  Correction: Two-line correction method; continuum-source correction method; background correction based on Zeeman effect; background correction based on source self-reversal.
- *Chemical interference:* various chemical processes occurring during atomization that alter the absorption characteristics of the analyte.
  Causes: Formation of compounds of low volatility, dissociation and association reactions lead to inter-conversion of the constituents; ionization leads to a significant concentration of electrons.

Cold Vapor Atomic Fluorescent Spectrometry (CVAFS) (Skook, Holler, and Neiman 1998)

CVAFS is very similar to CVAAS, the only difference being the method of detection. In it, the fluorescence of the excited mercury atoms is measured. In theory, atomic fluorescence measurements need no monochromator or filters since the emitted radiation is that of a single element and will excite only atoms of that element. Thus a non-dispersive system could be made up of only a source, an atomizer, and a detector. If it is possible to have a pure source and an atomizer with the least background interference, this system could be advantageous in the following ways:

- it would be a simple and low-cost system;
- it could be adapted for analyses of different elements just by replacing the source; and
- it achieves a high sensitivity since it can achieve high-energy throughput and ability to collect energy from multiple lines simultaneously.

The type and magnitude of interference in CVAFS system are very similar to that of CVAAS.

Inductively Coupled Plasma-Mass Spectrometry (ICP-MS) (Skoog, Holler, and Neiman 1998)

Because of its low detection limit, high degree of selectivity, and reasonably good precision and accuracy, ICP-MS has become popular in elemental analyses, including that of mercury.

The torch serves as an atomizer and ionizer. The physical state of the sample determines the sample introduction technique. Since most of the samples after preparation are in the aqueous state, conventional sample introduction techniques like flow injection, HPLC, or ultrasonic nebulizer are used.

The positive metal ions produced in a conventional ICP torch are sampled through a differentially pumped interface linked to a quadrupole mass spectrometer. The spectra thus produced consist of a simple series of isotope peaks for each of the elements present in the sample. These spectra are used for qualitative determination of the elements present and for the quantitative measurement of their amounts. Usually the latter are based on the calibration curves in which the ratio of the ion count for the analyte to the count of an internal standard is plotted as a function of concentration.

Sample Treatment

Samples like fish tissues and sediments, and some seawater, which have very complex matrices regarding other chemical species, need special treatment procedures to bring the sample into a form and the mercury into a constituent suitable for a certain separation and detection system.

Almost all methods for methylmercury in biological samples have focused specifically on the mono methyl form (MeHg) for reasons mentioned earlier. An estimated 85% of all MeHg in aquatic system resides in fish (Horvat, Bloom, and Liang, 1993). Most of the methods for mercury analyses in biological samples like fish tissue are based on or are some variation of the Westoo Method (Westoo 1966).

Westoo (1966) used GLC to isolate the methyl mercuric chloride (the final form of the mercurial in the above treatment process) and an ECD for detection and quantification.

Over the years, the original Westoo method of sample treatment has undergone several modifications. Some of the modifications are discussed by Mushak (1987). The basic Westoo procedure, however, continued to be used for method validations in many studies.

Bloom (1989) reportedly achieved a detection limit of 0.6 pg of MeHg (as Hg) or 0.003 ng/L for a 200-mL sample. In his procedure, a fish tissue is digested with potassium hydroxide (KOH)/ methanol for 24–48 hours at room temperature, or for one hour in an ultrasonic bath. The digest is then diluted with methanol and then neutralized buffered with acetic acid. The mercury in treated sample is then converted to the ethyl derivative. Sodium tetraethylborate ($NaB(CH_2CH_3)_4$) is used as the ethylation reagent in this procedure.

The analyte is then purged with N_2 onto a Carbotrap column, then thermally desorped onto a cryogenic GC column. Elute species are ramped-heated and a carrier gas passed through the thermal decomposer. Detection and quantification are then achieved using peak area/retention time and CVAFS.

In a recent study, Palmieri and Leonel (2000) reported a detection limit of 0.4 pg of MeHg (as Hg). In the study, freeze-dried fish powder and fresh fish tissue were analyzed for methylmercury. The samples were digested with KOH/methanol and acidified prior to extraction with methlyene chloride. Methylmercury was then back-extracted from the organic phase into water. An aliquot was then derivatized with sodium tetraphenylborate and then extracted with toluene. The organic phase containing MePhHg was injected into a GC column which was on-line with MIP-AED (Microwave Induced Plasma Atomic Emission Detection).

Although methods in more recent studies are essentially rectified versions of the earlier methods, improved detection limits have been achieved. Table A-2 summarizes the methods used to analyze different samples and the associated detection limits.

Before going on to discuss other environmental samples, it should be noted that evidently all of the above methods are not suitable for routine monitoring purposes. It would be appropriate to mention efforts directed for this purpose.

Horvat, Liang, and Bloom (1993) compared the different isolation methods for sediment and water samples in two different studies. In Part 1 (sediments), steam distillation, alkaline digestion, HCl leaching, and ion exchange methods were compared (Horvat, Bloom, and Liang 1993). In Part 2 (water), solvent extraction and distillation were compared. In both studies, the isolation was followed by aqueous phase ethylation, pre-collection on Carbotrap, isothermal GC, and CVAFS (Horvat, Liang, and Bloom 1993).

Table A-2. Summary of the Analytical Methods to Determine Mercury Species in Various Samples.

Methods (sample type)	Detection Limits (unless specified otherwise)	References
Automated system utilizing alkaline digestion and CVAAS (water)	1 ng/L as Hg	Goulden and Anthony 1980
CVAAS after preconcentration on a chelating resin (natural water)	0.2-5000 ng/L for 20 mL sample	Minagawa, Takizawa, and Kifune 1980
Two-stage gold amalgamation technique, CVAAS (seawater)	0.1 ng/L Hg	Bloom and Crecelius 1983
Preconcentration on Carbotrap/thermal desorption on a U-tube GC column and CVAAS (air)	0.3 pg for mercury and dimethylmercury, 0.4 pg for diethyl mercury, and 2.0 pg for monomethylmercury	Bloom and Fitzgerald 1988
Aqueous phase ethylation, Cryogenic GC, CVAFS (natural surface waters)	~0.6 pg as Hg or 0.003 ng/L for a 200 mL sample	Bloom 1989
Distillation followed by aqueous phase ethylation, precollection on carbotrap, isothermal GC and CVAFS (sediments)	0.001 ng/g MeHg as Hg for 100 mg	Horvat, Bloom, and Liang 1993
Distillation followed by aqueous phase ethylation, precollection on carbotrap, isothermal GC and CVAFS (water)	0.006 ng/L MeHg for 50-mL sample	Horvat, Liang, and Bloom 1993
Aqueous phase ethylation, room temperature precollection and GC/CVAFS (reagent blank)	0.6 pg for monomethylmercury and 1.3 pg for Hg(II)	Liang, Horvat, and Bloom 1994
Extraction, GC-AED (biological marine materials)	0.8 pg MeHgCl as Hg	Donais *et al.* 1996
Gas/Liquid separation using N_2 as carrier, gold amalgamation, an automatic flow injection system using commercial CVAAS (fish, hair, natural water)	5.3 ng/L Hg (absolute detection limit of 30 pg)	Canela, Jardim, and Rohwedder 1996
Steam distillation, GC, AFS (natural water)	0.024 ng/L for a 50 mL sample	Bowles and Apte 1998
Capillary electrophoresis coupled on-line with quadrupole and double-focusing ICP MS (standard solutions)	4 μg/L as Hg for inorganic mercury, 7 μg/L as Hg for organic mercury	da Rocha *et al.* 2000
Desorption of GDM collected in the KCL coated denuders, preconcentration on Au trap, CVAFS (gaseous divalent mercury in air)	3 pg	Xinbin *et al.* 2000
Capillary electrophoresis ICP-MS with microconcentric nebulization (contact lens solution)	0.08-0.17 μg/mL as Hg for various Hg species	Lee and Jiang 2000
GC-after derivatization with sodium tetra-ethylborate, MIP-AES (fish)	0.4 pg as Hg or quantification limit of 0.6 μg/g and 0.1 μg/g of MeHg (as Hg) for 0.08 g of freeze-dried fish powder and 0.5 g of fresh samples, respectively	Palmieri and Leonel 2000
Potentiometric stripping analysis using gold electrodes from recordable CDs (oceanic and tap water)	0.25 ng/mL Hg^0	Richter *et al.* 2000
Bulk optode membrane for determination of Hg(II) (amalgam of odontologic use)	0.11 ng/L Hg(II)	Sanchez-Pedreno *et al.* 2000
Cavity ringdown laser absorption spectrometer (air)	0.50 ng/m^3	Spuler *et al.* 2000
Water vapor distillation, HPLC-ICP MS coupling technique with ultrasonic nebulization (sediment, soil leaves)	0.015 μg/kg for MeHg	Wilken and Falter 1998

For sediment samples, distillation was found superior, as it was able to avoid matrix effects, was consistent, and obtained high recoveries (95 ± 4%). HCl leaching was ineffective in releasing MeHg quantitatively. TOC seemed to affect the recovery of MeHg. KOH-methanol digestion could cause matrix interference when sulphide-rich humic sediment is to be analyzed.

Also distillation was preferred over solvent extraction for water samples. It gave consistent, high recoveries and was able to achieve better detection limits. Solvent extraction produced low and nonreproducible recoveries from anoxic and humic-rich water samples as well as from particulate and humic-rich water samples.

Distillation has been used in many studies for the same reasons enumerated in the table. A recent study by Bowles and Apte (1998) confirms the above assertion. Steam distillation followed by aqueous phase ethylation, isothermal GC, and CVAFS detection was able to achieve a detection limit of 0.024 ng/L for a 50-mL natural water sample. Recoveries of about 100% of MeHg in fresh and estuarine water samples were also achieved. Addition of ammonium pyrrolidine dithiocarbonate (APDC) to seawater improved the poor recovery to about 85%. The main advantages of this method are the high sample throughput (at least 4/hr without any compromise in the analytical performance) and the capacity to handle sample sizes up to 100 mL. The study claims the absence of artifactual MeHg formation.

Hintelmann (1999) investigated artifactual MeHg formation from inorganic Hg^{2+}. Alkaline digestion, acidic leaching, and atmospheric pressure distillation were found to form $CH_3{}^{200}Hg^+$ from $^{200}Hg^{2+}$ added to CRMs. Of the added stable inorganic Hg isotope, 0.006–0.05% was converted to MeHg. Only leaching using a $H_2SO_4/KBr/CuSO_4$ mixture did not cause any artefacts, which implies that analysis of certain sediment samples where the MeHg content is less than 1% will be affected significantly by artifactual formation. For more accurate measurement, a species specific isotope addition (SSIA) was suggested.

Falter, Hintelmann, and Quevauviller (1999) adequately summarized the problem of artifactual formation. It is of utmost importance in mercury analyses not only for accurate quantification but also for fate studies.

As it is evidenced in all the methods discussed, chromatography, especially GC, has been used almost exclusively. This is because chromatographic separation is species specific and is readily coupled with spectrometric detection systems. The effectiveness of chromatographic separation for mercury speciation depends on the column packing materials, the carrier, and also the orientation of the columns. The most commonly used chromatography for mercury analyses are GC, GLC, and High Performance Liquid Chromatography (HPLC). The performances of different GLC column packings have been discussed by Mushak (1987). Robertson *et al.* (EPRI 1987) discusses the separation schemes for organo-mercury compounds using normal phase HPLC and silica gel column chromatography.

Although improved chromatography of some sort continues to dominate the studies in mercury analyses, an alternative method of gold (Au) amalgamation has also been gaining substantial attention. This method, however, is not species specific. It has been used to measure the total inorganic Hg. Mercury's strong affinity to gold allows it to be separated from other metal species present in the sample.

Two-stage Au amalgamation has been used successfully to remove interfering substances from the mercury vapor before it enters a detector. Fitzgerald and Gill (1979) conducted mercury analyses using a two-stage Au amalgamation gas sampling train with detection of the eluting elemental Hg by flameless atomic absorption. The analyses achieved a coefficient of variation of 4% for 0.5 ng Hg and detection limit of 0.06 ng of Hg. The above method was refined by Bloom and Crecelius (1983) and applied to seawater samples. This refined two-stage Au amalgamation pre-collection coupled with CVAA achieved a detection limit of 0.1 ng/L Hg. The concentrations in Puget Sound measured using this method were in the 0.5 ng/L range, significantly less than the earlier reported values of 10–20 ng/L. This study also investigated three oxidation methods usually utilized prior to amalgamation. Cold-oxidation using BrCl was found to be better than the rest. The permanganate hot-oxidation produced large variability and high detection limit for natural samples. Acid-storage and cold-oxidation by BrCl gave comparable results. Because of the possibility of contamination due to strongly bound organo-mercurials, it was recommended that the BrCl technique be used.

Yet another isolation/separation technique is ion-exchange. It has been used extensively in the analyses of other metals. In the case of mercury, however, it has not yet received much attention. It may be because the existing methods are deemed adequate. Nonetheless, it is worth a mention. An ion-exchange technique is based on its selectivity for the free-metal ion species (Hg^{2+} in this case) in the presence of typical monomeric ligands. The theory behind the method is elaborated by Fortin and Campbell (1998). Separation of MeHg by an anion exchange column after acidification with 6 M HCl and quatification by CVAAS was applied to water and soil samples by Padberg *et al.* (1994). Used on its own, this method may lead to an unreliable separation of MeHg in water and soil samples due to matrix effects. Especially the content of humic substances may cause a variation in the results depending on incomplete recovery in a derivatization step. With further development and refinement, this method may be applied as successfully to Hg species separation as it has been applied to other free-metal ion measurements.

Canela, Jardim, and Rohwedder (1996) developed an automatic method based on gold amalgamation, an automatic flow injection (FI) system, and commercial CVAAS to determine mercury in various environmental samples. The procedure was used to analyze fish, hair and natural water samples. A detection limit of 5.3 ng/L Hg (30 pg absolute) for the optimized system using 5.7 mL (3 injections of 1900 µL of the sample) with an analytical throughput of 6 samples/hr and reproducibility of 5%.

Detectors

The detectors that have been used in mercury analyses could be broadly categorized as following:

1. Electron Capture Detection (ECD),
2. Mass Spectrometry (MS),
3. Atomic Absorption/Emission Spectrometry (AA/ES), and
4. Atomic Fluorescence Spectrometry (AFS).

The above detectors have been used with various modifications. Rather than discussing each system of detection, a summary table has been produced below for comparison. The table also includes the separation technique used prior to detection.

Table A-3. Detector Systems for Mercury Analyses.

Detector System	Advantages and Disadvantages	Reference (Analytical Procedure)
ECD	Easy to use and very sensitive, but not species specific. It is sensitive to all electron capturing agents (not element specific) and thus not so accurate.	Westoo 1966 (GLC)
Cold vapor atomic absorption spectrometry (CVAAS)	Similar to CVAFS and very popular. Detection system of choice in EPA method 1631. Particularly suited for mercury analysis. Easy to use and convenient for routine uses. It can achieve high accuracy if the sample treatment is correctly chosen. It is not species specific.	Canela, Jardim, and Rohwedder 1996 (Au amalgamation-automatic FI); Bloom and Crecelius 1983 (Au amalgamation)
Cold vapor atomic fluorescence spectrometry (CVAFS)	Element specific, but not species specific. Particularly suited for mercury analysis. Easy to use and convenient for routine uses. It can achieve high accuracy if the sample treatment is correctly chosen. Has not been popular due to the wide success of CVAAS, which predates it by about a decade.	Bloom and Fitzgerald 1998 (GC); Horvat, Liand, and Bloom 1993 (collection on carbotrap and isothermal GC); Bloom 1989 (Cryogenic GC); Liang *et al.* 1994 (Isothermal GC, Pyrolytic Decomposition); Bowles and Apte 1998 (Steam Distillation)
Microwave Induced Plasma Atomic Emission Spectrometry (MIP) AES	Not used widely because it is not available from manufacturers. Its instrumentation is not as simple as CVAAS and CVAFS. Thus it requires proper training. Furthermore, it is more expensive than the two detectors mentioned previously. For this region, it is suited for laboratories for scientific investigations.	Donais *et al.* 1996 (capillary GC); EPRI 1987; Palmieri and Leonel 2000 (Derivatization with sodium tetraethyl borate and GC)
Inductively Coupled Plasma Mass Spectrometry (ICP) MS	Easily adaptable for multi-element analyses. A low detection limit of 0.015 ng/g for methylmercury can be achieved. Allows isotope-specific analyses in a very short time. The features are suited for scientific analyses like speciation. Expensive compared to the above detectors and needs proper training to use. Not suited for routine monitoring purposes.	Wilken and Falter 1998 (Ultrasonic nebulization, HPLC); Falter 1999

The major concern in a detection system has been the interference caused by species other than mercury. Bloom (1989) added a variety of chemical species to methylmercury-spiked tap water (0.54 ng/L Hg as CH_3HgCl) prior to ethylation, to determine potential interference. He found that of the common environmental anions, chloride caused strong interference resulting in reduced recovery of methyl ethyl Hg to 95% at 200 mg Cl^-/L and to less than 10% at 5000 mg Cl^-/L. Also at 10 mg Cl^-/L, a colloidal precipitate of Fe (III) hydroxide was observed and methylmercury recovery was reduced to 50%. Most often, the interference can be avoided by proper sample preparation and isolation techniques for accurate measurement of mercury species.

Conclusion

Analytical methods for determining mercury species in environmental samples are adequate. Unless alternative methods to the existing conventional ones are sufficiently developed, effort needs to be directed at curbing the weaknesses in the existing methods. Most of these analytical methods have been able to achieve detection limits, analytical precision, and accuracy adequate for the majority of the environmental samples. What remains to be done is the simplification of the existing methods so as to render them economically (both timewise and moneywise) suitable for routine monitoring analyses. Much has already been achieved in developing clean conditions that seemed to have affected the reliability of the data prior to the mid-1980s. The Au amalgamation and ion exchange methods of isolation have potential for improvement to be cost- and time-effective alternatives to GC and need further study.

The understanding of the fate of mercury is as equally contingent on the availability of reliable analytical methods as the alleviation of mercury-pollution is contingent on the understanding of the fate of mercury in the environment. The investigations on artifact formation facilitated by the species-specific isotope addition method are already shedding light on the methylation of inorganic mercury.

APPENDIX B

Recent Studies on the Bioaccumulation of Methylmercury in Lower Trophic Forms

Test Method or Study Description	Media Concentration: Water, Sediment	Species of Hg	Study Organism	Tissue Concentration of Test Organism	Water Quality Characteristics Evaluated	Duration of Exposure or Study	Correlation	Notes	References
Field study	Preflood lake water 0.5 ng/L; Flood conditions lake water 2.0 ng/L	Total Hg, MeHg	Insects of 7+ taxa	Post flood concs. Insects grouped into feeding guilds: Predators 189 ng/g; Collector/ shredders 71.5 ng/g	–	Concentrations of MeHg assessed for 1 year post flooding.	Increase in MeHg seen after flooding	Concentrations of MeHg presented for pre- and post flood conditions for 11+ genera of insects in a small wetland lake.	Hall, Rosenberg, and Wiens 1998
Rate of methylation tracked using ^{203}Hg	–	MeHg	Periphyton	–	–	2–4 hour incubation	–	Novel site of methylation and its relationship to photosynthesis.	Cleckner 1999
Rate of methylation tracked using ^{203}Hg	–	MeHg	*Salvinia* sp., *Eichornia* sp.	–	–	3 day incubation	–	Methylation sites, and role of bioturbation for delivering Hg to sediments.	Guimaraes *et al.* 1998
Mer-lux bioassay	–	$HgCl_3^-$ $HgCl_2$ $Hg(OH)_2$ $HgCL_4^{2-}$	*Escherichia coli*	–	–	100 min	DOC exerted inhibitory effect on methylating bacteria. Increasing Cl^- reduced bioavailability of inorganic Hg to bacteria	Bioavailability of various forms of Hg. Effect of DOC and NaCl.	Barkay, Gillman, and Turner 1997
Indoor microcosm	–	MeHg $HgCl_2$	*Chlorella vulgaris* *Daphnia magna* *Onchory-nchus mykiss* *(alevins)* *Gambusia affinis*	Concentrations reported for MeHg and inorganic Hg for all species	–	30 days exposure for BCF determin-ation for MeHg in *G. affinis*	–	Rates of transfer from trophic levels 1 through 4 for inorganic and organic Mercury species. MeHg BCFs calculated for *G. affinis* for direct route (2500) and direct route plus diet (27000) after 30 days exposure.	Boudou and Ribeyre 1997a, b

Test Method or Study Description	Media Concentration: Water, Sediment	Species of Hg	Study Organism	Tissue Concentration of Test Organism	Water Quality Characteristics Evaluated	Duration of Exposure or Study	Correlation	Notes	References
In vitro incubation	£4 ng/L range of concentrations used for uptake studies	MeHg Inorganic Hg	*Daphnia magna*	Concentrations of MeHg and inorganic Hg reported for uptake studies	Hardness, alkalinity, aquatic humus (AH)	24 hr–72 hr	Hardness, alkalinity no effect on uptake of MeHg. High AH increased total Hg in daphnids, but reduced bioavailability of MeHg.	Rates of uptake of MeHg and inorganic Hg determined for *D. magna* from water, and dietary sources. Effects of hardness, alkalinity, AH studied on rate of uptake.	Monson and Brezonik 1999
Field study	Water 0.00005 ppb	Unextracted MeHg (labile MeHg) Extracted MeHg (total MeHg) and total Hg	Plankton ($\geq$ 300 mm)	Plankton 10 ppb total MeHg Fish 100 ppb total MeHg	Acid neutralizing capacity (ANC), chlorophyll a, pH, DOC, color, P, SO_4	Spring, summer, fall for two consecutive years	Lack of significant correlations between plankton-Hg with ANC, pH, DOC, chlorophyll, P, SO_4	12 softwater lakes in NE Minnesota studied. Log BAFs determined for MeHg in plankton. Hg concentrations in plankton increased over growing season. However, MeHg declined in lake water over growing season. Log BAFs therefore variable. Authors differentiate between bioavailability of two types of MeHg.	Monson and Brezonik 1998
Field study	Unfiltered water – epilimnion 0.1–1.0 ng/L total Hg, <0.05 ng/L MeHg Unfiltered water – hypolimnion 1.2–3.0 ng/L total Hg 0.07–1.2 ng/L MeHg	Total Hg MeHg	Phytoplankton zooplankton, bluegill, shiners	Phytoplankton 9.1–9.6 ng/g MeHg Daphnia 100–186 ng/g MeHg Bluegill 265– 837 ng/g MeHg Shiner 287–483 ng/g MeHg	–	Three year study, 1993–1995 of bluegill and lake water. Two year study 1994–1995 of phytoplankton, zooplankton	–	Seasonal concentration changes in MeHg and total Hg determined in phytoplankton, zooplankton, and yearling mimic shiners. Fall destratification shown to be important time of entry of MeHg into food chain. Ranges of values given in this table for water and tissue concentrations. BMFs calculated for phytoplankton to Daphnia (11 and 19), Daphnia to shiners (2.6 and 2.7), Daphnia to bluegill (2.7 and 2.7).	Herrin *et al.* 1998

Test Method or Study Description	Media Concentration: Water, Sediment	Species of Hg	Study Organism	Tissue Concentration of Test Organism	Water Quality Characteristics Evaluated	Duration of Exposure or Study	Correlation	Notes	References
Field study	Average measure-ments (in ng/L) of 15 lakes: total Hg 1.94; dissolved Hg 1.71; total MeHg 0.28; dissolved MeHg 0.25; total HgII (Hg–MeHg – Hg^0) 1.66; dissolved HgII 1.46; dissolved gaseous Hg (Hg^0) 21 pg/L	See previous column	Microseston, zooplankton, small fish	Microseston 34 ng/g MeHg d.w. Zooplankton 53 MeHg ng/g d.w. Small fish 485 MeHg ng/g d.w.	DOC, pH, DIC, SPM, Chlor a, cond., Ca, Mg, Na, Fe, Mn, SO_4, Cl	–	Inverse relationship between pH and mercury concentration for fish, microseston. Stronger relationship than weak positive relationship between DOC and MeHg concentration in fish, and microseston.	15 northern Wisconsin lakes studied. Authors suggest: 1) passive uptake does not control bioaccum-ulation at base of aquatic food web; 2) correlation with pH and DOC reflect supply and bioavailability of Me Hg to lower trophic levels; and 3) MeHg levels in higher trophic levels reflect uptake at lower trophic levels, and other factors such as diet and growth.	Watras *et al.* 1998

Test Method or Study Description	Media Concentration: Water, Sediment	Species of Hg	Study Organism	Tissue Concentration of Test Organism	Water Quality Characteristics Evaluated	Duration of Exposure or Study	Correlation	Notes	References
Indoor microcosm	5-point concentration range set up for exposure of test organisms to sediment, or the water column	MeHg Inorganic mercury	Asiatic clam, *Corbicula fluminea*	MeHg and inorganic mercury determined and reported for whole bodies of clams and in individual organs.	–	14 day exposure	–	MeHg more rapidly bioaccumulated than did inorganic Hg from either water or sediment. Bioaccumulation was an order of magnitude higher for MeHg from water column than from sediment. Mantle and foot showed greatest affinity for MeHg, visceral mass showed affinity for inorganic Hg after exposure to the two forms of Hg.	Inza *et al.* 1997
Indoor microcosm	Dietary source of MeHg presented to test organisms	MeHg	Crayfish, *Astacus astacus*	MeHg determined and reported for individual organs and tissues	–	5, 10, 15 day exposures	–	Transfer rate of MeHg from Asiatic clam to *A. astacus* was 16%. Trophic uptake of MeHg results in high levels of bioaccumulation in the green gland and tail muscle of *A. astacus*, with an obvious risk of transfer to predators.	Simon , Ribeyre, and Boudou 2000
In vitro incubation	Sediment spiked with 95 ng/g d.w. MeHg	MeHg	Oligochaete worm, *Lumbricus variegatus*	MeHg determined in worm tissue	TOC, Se	14 day exposure	Uptake of MeHg lower in sediments with high organic carbon concentration. Addition of 2.5mg/kg Se to sediment lowered MeHg in test organism by 25%, 15mg/kg Se lowered MeHg by 75%, and 50mg/kg Se lowered MeHg by 86% as compared to controls.	Accumulation of MeHg from sediment investigated in an oligochaete worm.	Nuutinen and Kukkonen 1998

APPENDIX C

Mercury Accumulation by Piscivorous Fish and Birds

Species	Mercury Conc.[a] (µg/g fw)	Tissue	Prey Species and Conc. (µg/g fw)	Water Conc.[a] ng/L	Sampling Location	Water Quality Characteristics	Comments	Reference
Piscivorous Fish								
Oncorhynchus mykiss	1.84 MeHg[b]	muscle	*Retropinna retropinna*, 0.4 MeHg[b]; *Gobiomorphus cotidianus*, 0.2 MeHg[b]	4.5 8±0.82 Hg, 0.36±0.42 MeHg	New Zealand, Lake Rotomahana	pH = 6.7	clean techniques used for water samples	Kim and Burggraaf 1999
Oncorhynchus mykiss	0.6 MeHg[b]	muscle	*Retropinna retropinna*, 0.12 MeHg[b]; *Gobiomorphus cotidianus*, 0.15 MeHg[b]; *Paranephrops planifrons*, 0.15 MeHgb	1.9±1.05 Hg; 0.51±0.16 MeHg	New Zealand, Lake Okaro	pH = 6.8	clean techniques used for water samples	Kim and Burggraaf 1999
Oncorhynchus mykiss	0.5 MeHg[b]	muscle	*Retropinna retropinna*, 0.05 MeHg[b]; *Gobiomorphus cotidianus*, 0.5 MeHg[b]	1.47±0.57 Hg; 0.23±0.15 MeHg	New Zealand, Lake Rotorua	pH = 7.0	clean techniques used for water samples	Kim and Burggraaf 1999
Oncorhynchus mykiss	0.2 MeHgb	muscle	*Retropinna retropinna*, 0.02 MeHg[b]; *Gobiomorphus cotidianus*, 0.05 MeHg[b]; *Paranephrops planifrons*, 0.02 MeHg[b]	0.92±0.62 Hg; 0.09±0.06 MeHg	New Zealand, Lake Okareka	pH = 6.9	clean techniques used for water samples	Kim and Burggraaf 1999
Oncorhynchus mykiss	0.2 MeHg[b]	muscle	*Retropinna retropinna*, 0.02 MeHg[b]; *Gobiomorphus cotidianus*, 0.05 MeHg[b]; *Paranephrops planifrons*, 0.02 MeHg[b]	0.57±0.23 Hg; 0.21±0.07 MeHg	New Zealand, Lake Tarawera	pH = 7.5	clean techniques used for water samples	Kim and Burggraaf 1999
Micropterus salmoides	0.26–0.47	muscle	*Lepomis macrochirus*, 0.11–0.14	–	Connecticut, Pickerel Lake	impoundment	fish ages varied from 2–5 years	Neumann and Ward 1999
Esox niger	0.24–0.44	muscle	*Lepomis macrochirus 0.11–0.14*	–	Connecticut, Pickerel Lake	impoundment	fish ages varied from 2–5 years	Neumann and Ward 1999
Micropterus salmoides	0.33–0.89	muscle	*Lepomis macrochirus 0.17–0.28*	–	Connecticut, Lake Lillinonah		fish ages varied from 2–5 years	Neumann and Ward 1999
Micropterus dolomieu	0.39–0.91	muscle	*Lepomis macrochirus 0.17–0.28*	–	Connecticut, Lake Lillinonah		fish ages varied from 2–5 years	Neumann and Ward 1999

Species	Mercury Conc.[a] (µg/g fw)	Tissue	Prey Species and Conc. (µg/g fw)	Water Conc.[a] ng/L	Sampling Location	Water Quality Characteristics	Comments	Reference
Micropterus salmoides	0.11–0.52	muscle	*Lepomis macrochirus 0.28–0.08*	–	Oregon, Dorena Reservoir	pH = 7.7	fish ages varied from 2–5 years; sediment conc. 0.18±0.05 µg/gdw	Park and Curtis 1997
Micropterus salmoides	0.05–1.1 (0.39 avg)	muscle	–	–	Massachusetts, 24 least impacted lakes	pH = 4.9–10.5, DOC from <MDL to 58.8 mg/L	only largemouth bass longer than 30.5 cm selected; sediment conc. ranged from 0.008–0.425 µg/gdw	Rose *et al.*1999
Perca flavescens	0.01–0.75 (0.31 avg)	muscle	–	–	Massachusetts, 24 least impacted lakes	pH = 4.9-10.5, DOC from <MDL to 58.8 mg/L	fish selected were limited to 20–25 cm in length; sediment conc. ranged from 0.008–0.425 µg/gdw	Rose *et al.* 1999
Perca flavescens	0.13	muscle	–	–	Ontario, Canada, Plastic Lake	pH = 5.6	sediment conc. <5 ng/g dw	Scheuhammer and Graham 1999
Ambloplites rupestris	0.22	muscle	–	–	Ontario, Canada, Plastic Lake a	pH = 5.6	sediment conc. <5 ng/g dw	Scheuhammer and Graham 1999
Leopmis gibbosus	0.12	muscle	–	–	Ontario, Canada, Plastic Lake	pH = 5.6	sediment conc. <5 ng/g dw	Scheuhammer and Graham 1999
Leopmis gibbosus	0.04	muscle	–	–	Ontario, Canada, Blue Chalk Lake	pH = 6.9	sediment conc. <5 ng/g dw	Scheuhammer and Graham 1999
Micropterus salmoides	0.244– 1.084	muscle	–	–	Connecticut, mean data from 46 lakes	pH = 6.8–8.6	Hg conc. Was higher in reservoirs than drainage lakes	Hanten *et al.* 1998
Micropterus salmoides	0.33– 0.5	muscle	0.33–0.5	–	Oregon, Dorena Reservoir	pH = 7.7	Mean data from fish of average ages 4.8–6.0 years	Foster, Drake, and DiDomenico 2000

Species	Mercury Conc.[a] (µg/g fw)	Tissue	Prey Species and Conc. (µg/g fw)	Water Conc.[a] ng/L	Sampling Location	Water Quality Characteristics	Comments	Reference
Piscivorous Birds	µg/mL µg/gfw µg/gdw	blood tissue feathers						
Gavia immer; adults	0.93–4.26	blood	*Perca flavescens, Lepomis gibbosus, Ambloplites rupestris*; 0.05-0.43	–	Ontario, Canada, 24 lakes	pH = 6.23–8.23 DOC = 1.26–5.13 ppm	94% MeHg in prey fish	Scheuhammer *et al.* 1998
Gavia immer; chicks	0.044–0.609	blood	*Perca flavescens, Lepomis gibbosus, Ambloplites rupestris*; 0.05-0.43	–	Ontario, Canada, 24 lakes	pH = 6.23–8.23 DOC = 1.26–5.13 ppm	94% MeHg in prey fish	Scheuhammer *et al.* 1998
Gavia immer; adults	7.65–21.0	feathers	*Perca flavescens, Lepomis gibbosus, Ambloplites rupestris*; 0.05-0.43	–	Ontario, Canada, 24 lakes	pH = 6.23–8.23 DOC = 1.26–5.13 ppm	94% MeHg in prey fish	Scheuhammer *et al.* 1998
Gavia immer; chicks	1.37–3.35	feathers	*Perca flavescens, Lepomis gibbosus, Ambloplites rupestris* 0.05-0.43	–	Ontario, Canada, 24 lakes	pH = 6.23–8.23 DOC=1.26–5.13 ppm	94% MeHg in prey fish	Scheuhammer *et al.* 1998
Gavia immer; chicks	0.06–0.4[b]	blood	–	–	Wisconsisn, 45 lakes	pH = 5.0–8.5	acid neutralizing capacity and lake area improved curve fits	Meyer *et al.* 1998
Gavia immer; adults	0.8–3.2[b]	blood	–	–	Wisconisn, 45 lakes	pH = 5.0–9.5	Additional water chemistry variables did not improve curve fits	Meyer *et al.* 1998
Pandion haliaetus; 28–35 day-old chick	3.26– 7.40	feathers	*Perca flavescens* (4–5 years old); 0.05–0.10	–	Ontario, Canada, St. Mary's River, Georgian Bay, and Kawartha Lakes	–		Hughes, Ewins, and Clark 1997
Pandion haliaetus	0.5– 3.4a[b]	blood	*Esox lucius, Stizostedion vitreum, Coregonus sp., Catastomus sp;* 0.57–2.44	–	Ontario, Canada, La Grande Reservoirs	–		DesGranges *et al.* 1998

Species	Mercury Conc.[a] (µg/g fw)	Tissue	Prey Species and Conc. (µg/g fw)	Water Conc.[a] ng/L	Sampling Location	Water Quality Characteristics	Comments	Reference
Pandion haliaetus; 40-day-old chick	37.3	feathers	*Esox lucius, Stizostedion vitreum, Coregonus sp., Catastomus sp.*; 0.57–2.44	–	Ontario, Canada, La Grande Reservoirs	–		DesGranges *et al.* 1998
Ardea herodias; 42-day-old chick	0.87– 2.02	feathers	*Dorosoma* sp., *Cyprinus carpio, Ictalurus punctatus, Micrpterus* sp; 0.002–0.017	–	Oak Ridge, Tennessee, two colonies on the Oak Ridge Reservation and two offsite colonies	–		Halbrook, Brewer, and Beuhler 1999
Phalacrocorax auritus; 17–22 day-old nestlings	0.22–0.50 0.30–0.51 0.12–0.23 1.4–10.0	blood liver muscle feathers	*Morone chrysops*, 0.21; *Dorosoma petenense*, 0.13; *Dorosoma cepedianum* 0.05	0.1–0.7	New Mexico, Elephant Butte and Caballo reservoirs	–	relatively uncontaminated reservoir	Caldwell, Arnold, and Gould 1999
Phalacrocorax auritus	0.13–0.46	eggs	*Morone chrysops*, 0.21; *Dorosoma petenense*, 0.13; *Dorosoma cepedianum* 0.05	0.1–0.7	New Mexico, Elephant Butte and Caballo reservoirs	–	relatively uncontaminated reservoir	Caldwell, Arnold, and Gould 1999
Ardea herodias young	1.3 1.39–1.42 1.15 0.30–0.36 1.97–2.23	blood liver kidney brain feathers	*Lavinia exilicauda* 0.46–0.87	22–81 (0.1–1 MeHg)	California, Clear Lake	–	sediment Hg conc. 40–370 mg/kg	Wolfe and Norman 1998
Phalacrocorax auritus young	1.91–2.94 0.54–0.72 2.95–4.05	liver brain feathers	*Lavinia exilicauda* 0.31–0.38	22-81 (0.1–1 MeHg)	California, Clear Lake	–	sediment Hg conc. 40–370 mg/kg	Wolfe and Norman 1998
Piscivorous Mammals								
Procyon lotor	0.4 3.29 0.67 21.97 10.22	blood liver brain fur claws	–	22-81 (0.1–1 MeHg)	California, Clear Lake	–	sediment Hg conc. 40–370 mg/kg, conc. from animals collected closest to source (0.1–0.8 km)	Wolfe and Norman 1998

Species	Mercury Conc.[a] (μg/g fw)	Tissue	Prey Species and Conc. (μg/g fw)	Water Conc.[a] ng/L	Sampling Location	Water Quality Characteristics	Comments	Reference
Mustela vison	0.08–0.63 0.28–1.13 0.44–10.1 0.41–1.3 0.12–7.1 2.13–21.5 2.28–34.1	blood muscle liver kidney brain fur claws	–	22-81 (0.1–1 MeHg)	California, Clear Lake	–	sediment Hg conc. 40–370 mg/kg, conc. ranges from animals collected either 11.3 or 22.5 km from source	Wolfe and Norman 1998

REFERENCES

Albanus, L., L. Frankenberg, C. Grant, U. von Hartmann, A. Jernelov, G. Nordberg, M. Rydalv, A. Schutz, and S. Skerfving. 1972. Toxicity for cats of methylmercury in contaminated fish from Swedish lakes and of methylmercuryhydroxide added to fish. *Environ. Res.* 5:425-442.

Alexander, G. R. 1977. Food of vertebrate predators on trout waters in north central lower Michigan. *Mich. Acad.* 10:181-195.

Ambe, M. and K. Suwabe. 1977. The preparation of standard solutions of mercury at the p.p.b. level. *Analytica Chimica Acta* 92:55-60.

Amyot, M., G. Mierle, D. R. S. Lean, and D. J. McQueen. 1997. Effect of solar radiation on the formation of dissolved gaseous mercury in temperate lakes. *Geochimica et Cosmochimica Acta* 61(5):975-987.

ATSDR (Agency for Toxic Substances and Disease Registry). 1989. *Toxicological Profile for Selected PCBs (Aroclor 1260, 1254, 1248, 1242, 1232, 1221, and 1060).* ATSDR/TP-89-25. Atlanta, GA, USA: U.S. Public Health Service.

ATSDR (Agency for Toxic Substances and Disease Registry). 1997. *Toxicological Profile for Mercury.* Atlanta, GA: U.S. Department of Health and Human Services.

Aulerich, R. J., R. K. Ringer, and S. Iwamoto. 1974. Effects of dietary mercury on mink. *Arch. Environ. Contam. Toxicol.* 2(1):43-51.

Bailer J. A. and J. T. Oris. 1997. Estimating inhibition concentrations for different response scales using generalized linear models. *Environ. Toxicol. Chem.* 16:1554-1559.

Barkay, T., M. Gillman, and R. R. Turner. 1997. Effects of dissolved organic carbon and salinity on bioavailability of mercury. *Appl. Environ. Microbiol.* 63:4267-4271.

Barr, J. F. 1986. *Population dynamics of the Common Loon (Gavia immer) associated with mercury-contaminated waters in northwestern Ontario.* Ottawa: Canadian Wildlife Service.

Bartlett, P. D., P. J. Craig, and S. F. Morton. 1977. Behavior of mercury species in isolated estuarine sediment samples, growth and decay of methylmercury during storage. *Nature* 267: 606-608.

Beijer, K. and A. Jernelov. 1979. Methylation of mercury in aquatic environments. Chap. 9 in *The Biogeochemistry of Mercury in the Environment.* Amsterdam: Elsevier/North-Holland Biomedical Press.

Benoit, J. M., C. C. Gilmour, R. P. Mason, G. S. Riedel, and G. F. Riedel. 1998. Behavior of mercury in the Patuxent River Estuary. *Biogeochemistry* 40:249-265.

Bidwell, J. R. and A. G. Heath. 1993. An *in situ* study of rock bass (*Ambloplites rupestris*) physiology: Effect of season and mercury contamination. *Hydrobiologia* 264:134-152.

Biester, H. and C. Scholtz. 1997. Determination of mercury binding forms in contaminated soils: mercury pyrolysis vs. sequential extractions. *Environ. Sci Technol.* 31:233-239.

Birge, W. J., J. A. Black, A. G. Westerman, and J. E. Hudson. 1979. The effects of mercury on reproduction of fish and amphibians. In *The biogeochemistry of mercury in the environment.* Amsterdam: Elsevier/North-Holland Biomedical Press.

Bleavins M. R. and R. J. Aulerich. 1981. Feed consumption and food passage time in mink (*Mustela vison*) and European ferrets (*Mustela putorius furo*). *Lab Anim. Sci.* 3:268-269.

Bloom, N.S. 1995. Trace metals & ultra-clean sample handling. *Environ. Lab.* 7:20-25.

Bloom, N.S. 1992. On the chemical form of mercury in edible fish and marine invertebrate tissue. *Can. J. Fish. Aquat. Sci.* 49:1010-1017.

Bloom, N. S. 1989. Determination of picogram levels of methylmercury by aqueous phase ethylation, followed by cryogenic gas chromatography with cold vapor atomic fluorescence detection. *Can. J. Fish. Aquat. Sci.* 46:1131-1140.

Bloom, N. S. and E. A. Crecelius. 1983. Determination of mercury in seawater at sub-nanogram per liter levels. *Mar. Chem.* 14:49-59.

Bloom, N. S., W. F. Fitzgerald. 1988. Determination of volatile mercury species at the picogram level by low temperature gas chromatography with cold vapor atomic fluorescence detection. *Analytica Chimica Acta* 208:151-161.

Bloom, N. S., C. J. Watras, and J. P. Hurley. 1991. Impact of Acidification on the Methylmercury Cycle of Remote Seepage Lakes. *Water, Air and Soil Science* 56:477-491.

Bloom, N. S., L. Liang, Z. Q. Xie, and S. S. Wang. 1994. *Mercury pollution: integration and synthesis,* edited by C. J. Watras and J. W. Huckabee. Palo Alto, CA: Electric Power Research Institute.

Bodaly, R. A., J. W. M. Rudd, R. J. P. Fudge, and C. A. Kelly. 1993. Mercury concentrations in fish related to size of remote Canadian Shield lakes. *Can. J. Fish. Aquat. Sci.* 50:980-987.

Borg, K., H. Wantorp, K. Erne, and E. Hanko. 1969. Alkylmercury poisoning in terrestrial Swedish wildlife. *Virtrevy.* 6.

Borg, K., K. Erne, E. Hanko, and H. Wanntorp. 1970. Experimental secondary methylmercury poisoning in the goshawk (*Accipiter g. gentilis L.*). *Environmental Pollution* 1:91-104.

Bortleson, G. C., S. E. Cox, M. D. Munn, R .J. Schumaker, and E. K. Block. 1994. Sediment-quality assessment of Franklin D. Roosevelt Lake and upstream reach of the Columbia River, Washington, 1992. Open-File Report 94-315. Tacoma, WA: U.S. Geological Survey.

Bothner, M. H., and D. E. Robertson. 1975. Mercury contamination of sea water samples stored in polyethylene containers. *Anal. Chem.* 47:592-595.

Bothner, M. H., R. A. Jahnke, L. Peterson, and R. Carpenter. 1980. Rate of mercury loss from contaminated estuarine sediments. *Geochim. Cosmochim. Acta* 44:273-285.

Boudou, A. and F. Ribeyre. 1997a. Aquatic toxicology: from the ecosystem to the cellular and molecular levels. *Environ. Health Perspect.* 105:21-35.

Boudou, A. and F. Ribeyre. 1997b. Mercury in the food web: accumulation and transfer mechanisms. *Metal Ions in Biological Systems* 34:289-319.

Boutron, C. F. 1990. A clean laboratory for ultralow concentration heavy metal analysis. *Fresenius J. Anal. Chem.* 337:482-491.

Bowles, K. C. and S. C. Apte. 1998. Determination of methylmercury in natural water samples by steam distillation and gas chromatography-atomic fluorescence spectrometry. *Anal. Chem.* 70:395-399.

Brumbaugh, W. G., D. P. Krabbenhoft, D. R. Helsel, J. G. Wiener, and K. R. Echols. 2001. *A national pilot study of mercury contamination of aquatic ecosystems along multiple gradients: Bioaccumulation in Fish.* Biological Sciences Report. USGS/BRD/BSR-2001-009.

Bukaveckas, P. A. and C. T. Driscoll. 1991. Effects of whole-lake base addition on the optical properties of three clearwater acidic lakes. *Can. J. Fish. Aquat. Sci.* 48:1030-1040.

Burmaster D. E., K. J. Lloyd, and K. M. Thompson. 1995. The need for new methods to back-calculate soil cleanup targets in interval and probabilistic cancer risk assessments. *Hum. Ecol. Risk Assess.* 1:89-100.

Cabana, G., A. Tremblay, J. Kalff, and J. B. Rassmussen. 1994. Pelagic food chain structure in Ontario lakes: a determinant of mercury levels in lake trout (*Salvelinus namaycush*). *Can. J. Fish Aquat. Sci.* 51:381-389.

Cai, Y., R. Jaffé, A. Alli, and R. D. Jones. 1996. Determination of organomercury compounds in aqueous samples by gas chromatography-atomic fluorescence spectrometry following solid-phase extraction. *Analytica Chimica Acta* 334:251-259.

Caldwell, C. A., M. A. Arnold, and W. R. Gould. 1999. Mercury distribution in blood, tissues, and feathers of double-crested cormorant nestlings from arid-lands reservoirs in south central New Mexico. *Arch. Environ. Contam. Toxicol.* 36:456-461.

Canela, M. C., W. F. Jardim, and J. J. R. Rohwedder. 1996. Automatic determination of mercury in samples of environmental interest. *J. Auto. Chem.* 18:193-198.

Carr, R. A. and P. E. Wilkniss. 1973. Mercury: Short-term storage of natural waters. *Environ. Sci. Technol.* 7:62-63.

Carron, J. and H. Agemian. 1977. Preservation of sub-p.p.b. levels of mercury in distilled and natural freshwaters. *Analytica Chimica Acta* 92:61-70.

Castilhos, Z. C. and E. D. Bidone. 2000. Hg biomagnification in the ichthyofauna of the Tapajós River region, Amazonia, Brazil. *Bull. Environ. Contam. Toxicol.* 64:693-700.

Cavalli, S. and N. Cardellicchio. 1995. Direct determination of seleno-amino acids in biological tissues by anion-exchange separation and electrochemical detection. *J. Chromatog.* 706(A): 429-436.

Chamberland, G., D. Belanger, and A. Dallaire. 1996. Urinary protein excretion of semi-domesticated mink in a chronic methylmercury study. *Journal of Toxicology and Environmental Health.* 47:285-297.

Charbonneau, S. M., I. C. Munro, E. A. Nera, R. F. Willes, T. Kuiper-Goodman, F. Iverson, C. A. Moodie, D. R. Stoltz, F. A. J. Armstrong, J. F. Uthe, and H. C. Grice. 1974. Subacute toxicity of methylmercury in the adult cat. *Toxicol. Appl. Pharm.* 27:569-581.

Cleckner, L. B. 1999. Mercury methylation in the Florida Everglades. *Limnol. Oceanogr.* 44:1815-1825.

Colman, J. A. and R. F. Breault. 2000. Sampling mercury at subnanogram per litre concentrations for load estimation in rivers. *Can. J. Fish. Aquat. Sci.* 57:1073-1079.

Connors, P. G., V. C. Anderlini, R.W. Risebrough, M. Gilbertson, and H. Hays. 1975. Investigations of heavy metals in common tern populations. *Can. Field. Nat.* 89:157-162.

Cope, W. G., J. G. Wiener, and R. G. Rada. 1990. Mercury accumulation in yellow perch in Wisconsin seepage lakes: relation to lake characteristics. *Environ. Toxicol. Chem.* 9: 931-940.

Conroy, N. and W. Keller. 1976. Geological factors affecting biological activity in precambrian shield lakes. *Canadian Mineralogist* 14:62-72.

Cronan, C. S. and C. L. Schofield. 1979. Aluminum leaching response to acid precipitation: effects of high-elevation watersheds in the Northeast. *Science* 204:304-306.

D'Itri, F.M. 1972. *The Environmental Mercury Problem.* Cleveland, OH: The Chemical Rubber Company.

Dansereau, M., N. Lariviere, D. Du Tremblay, and D. Belanger. 1999. Reproductive performance of two generations of female semidomesticated mink fed diets containing organic mercury contaminated freshwater fish. *Arch. Environ. Contam. Toxicol.* 36:221-226.

da Rocha, M. S., A. B. Soldado, E. Blanco-Gonzalez, and A. Sanz-Medel. 2000. Speciation of mercury compounds by capillary electrophoresis coupled on-line with quadrupole and double-focusing inductively coupled plasma mass spectrometry. *J. Anal. Atomic Spectrometry* 15: 513-518.

De Haan, H. 1992. Impacts of environmental changes on the biogeochemistry of aquatic humic substances. *Hydrobiologia* 229:59-71.

DesGranges, J. L., J. Rodrigue, B. Tardif, and M. Laperle. 1998. Mercury accumulation and biomagnification in ospreys (*Pandion haliaetus*) in the James Bay and Hudson Bay regions of Québec. *Arch. Environ. Contam. Toxicol.* 35:330-341.

Dillon, P. J., N. D. Yan, and H. H. Harvey. 1984. Acidic deposition: effects on aquatic ecosystems. *CRC Critical Reviews in Environmental Control.* 13:167-194.

Donais, M. K., P. C. Uden, M. M. Schantz, and S. A. Wise. 1996. Development, validation, and application of a method for quantification of methylmercury in biological marine materials using gas chromatography atomic emission detection. *Anal. Chem.* 68:3859-3866.

Driscoll, C. T., C. Yan, C. L. Schofield, R. Munson, and J. Holsapple. 1994. The mercury cycle and fish in the Adirondack lakes. *Environ. Sci. Technol.* 28:136A-143A.

Driscoll, C. T., V. Blette, C. Yan, C. L. Schofield, R. Munson, and J. Holsapple. 1995. The role of dissolved organic carbon in the chemistry and bioavailability of mercury in remote Adirondack lakes. *Water, Air, and Soil Pollut.* 80:499-508.

Eaton, R. D. P., D. C. Secord, and P. Hewitt. 1980. An experimental assessment of the toxic potential of mercury in ringed-seal liver for adult laboratory cats. *Toxicol. Appl. Pharmacol.* 55:514-521.

Effler, S. W., G. C. Schafran, and C. T. Driscoll. 1985. Partitioning light attenuation in an acidic lake. *Can. J. Fish. Aquat. Sci.* 42:1707-1711.

Eganhouse, R. P. 1975. *The measurement of total and organic mercury in marine sediments, organisms and waters.* El Segeindo: Southern California Coastal Water Research Project.

Eilers, J. M., G. E. Glass, K.E. Webster, and J. A. Rogalla. 1983. Hydrologic control of lake susceptibility to acidification. *Can. J. Fish. Aquat. Sci.* 40:1896-1904.

EPRI (Electric Power Research Institute). 1987. *Measurement of bioavailable mercury species in fresh water and sediments.* EPRI EA-5197, Project 2020-3, Final Report, June 1987.

Evers, D. C., J. D. Kaplan, M. W. Meyer, P. S. Reaman, W. E. Braselton, A. Major, N.Burgess, and A.M. Scheuhammer. 1998. Geographic trend in mercury measured in common loon feathers and blood. *Environmental Toxicology and Chemistry* 17:173-183.

Falandysz, J. 1986. Metals and organochlorines in adult and immature males of white-tailed eagle. *Environ. Conserv.* 13:69-70.

Falandysz, J., B. Jakuczun, and T. Mizera. 1988. Metals and organochlorines in four female white-tailed eagles. *Mar. Pollut. Bull.* 19(10):521-526.

Falter, R. 1999. Experimental study on the unintentional abiotic methylation of inorganic mercury during analysis. Part 1: Localisation of the compounds effecting the abiotic mercury methylation. *Chemosphere* 39:1051-1073.

Falter, R., H. Hintelmann, and P. Quevauviller. 1999. Conclusion of the workshop on "sources of error in methylmercury determination during sample preparation, derivatization and detection". *Chemosphere* 39:1039-1049.

Ferrari, C. P., A. L. Moreau, and C. F. Boutron. 2000. Clean conditions for the determina-tion of ultra-low levels of mercury in ice and snow samples. *Fresenius. J. Anal. Chem.* 366: 433-437.

Ferson S. 1995. Using approximate deconvolution to estimate cleanup targets in probabilistic risk analyses. In *Hydrocarbon Contaminated Soils*, Volume 5. Amherst, MA: Amherst Scientific Publishers. pp 245-254.

Ferson S. 1996. What Monte Carlo methods cannot do. *Hum. Ecol. Risk Assess.* 2:990-1007.

Fimreite, N. 1974. Mercury contamination of aquatic birds in northwestern Ontario. *J. Wildl. Manage.* 38(1):120-131.

Fimreite, N. 1971. *Effects of methylmercury on ring-necked pheasants, with special reference to reproduction.* Ottawa: Canadian Wildlife Service.

Fimreite, N. and Karstad, L. 1971. Effects of dietary methylmercury on red-tailed hawks. *J. Wildl. Manage.* 35:292-300.

Finley, M. T. and R. C. Stendall. 1978. Survival and reproductive success of black ducks fed methylmercury. *Environ. Poll.* 16:51-64.

Finley, M. T., W. H. Stickel, and R. E. Christensen. 1979. Mercury residues in tissues of dead and surviving birds fed methylmercury. *Bull. Environ. Contam. Toxicol.* 21:105-110.

Fitzgerald, W. F. and G. A. Gill. 1979. Subnanogram determination of mercury by two-stage gold amalgamation and gas phase detection applied to atmospheric analysis. *Anal. Chem.* 51:1714-1720.

Fitzgerald, W. F. and C. J. Watras. 1989. Mercury in surficial waters of rural Wisconsin lakes. *Sci Total Environ.* 87:223-232.

Fortin, C. and P. G. C. Campbell. 1998. An ion-exchange technique for free-metal ion measurements (Cd^{2+}, Zn^{2+}): Applications to complex aqueous media. *Intern. J. Environ. Anal. Chem.* 72:173-194.

Foster, E. P., D. L. Drake, and G. DiDomenico. 2000. Seasonal changes in tissue distribution of mercury in largemouth bass (Micropterus salmoides) from Dorena Reservoir, Oregon. *Arch. Environ. Contam. Toxicol.* 38:78-82.

Freeman, H. C. and D. A. Horne. 1973. Mercury in Canadian seals. *Bull. Environ. Contam. Toxicol.* 10(3):172-180.

Furness, R. W., J. L. Johnston, J. A. Love, and D. R. Thompson. 1989. Pollutant burdens and reproductive success of golden eagles *Aquila chrysaetos* exploiting marine and terrestrial food webs in Scotland. In *Raptors in the Modern World*, edited by B.U. Meyburg and R.D. Chancellor. Berlin: WWGBP.

Fyfe, R. W., R. W. Risebrough, and W. Walker. 1976. Pollutant effects on the reproduction of the prairie falcons and merlins of the Canadian prairies. *Can. Field Nat.* 90:346-355.

Fynn-Aikins, K., E. Gallagher, S. Ruessler, J. Wiebe, and T. S. Gross. 1998. An evaluation of methylmercury as an endocrine disruptor in largemouth bass. *SETAC 19th Annual Meeting.* Charlotte, N.C.

Ganther, H. E., C. Goudie, M. L. Sunde, M. J. Kopecky, P. Wagner, S. Oh, and W. G. Hoekstra. 1972. Selenium: Relation to decreased toxicity of methylmercury added to diets containing tuna. *Science* 175:1122-1124.

Gardiner, E. E. 1972. Differences between ducks, pheasants and chickens in tissue mercury retention, depletion and tolerance to increasing levels of dietary mercury. *Can. J. Anim. Sci.* 52:419-423.

Gilbert, M. M., J. Sprecher, L. W. Chang, and L. F. Meisner. 1983. Protective effect of vitamin E on genotoxicity of methylmercury. *J. Toxicol. Environ. Health.* 12:767-773.

Gilman, A. P., G. A. Fox, D. B. Peakall, S. M. Teeple, T. R. Carroll, and G. T. Haymes. 1977. Reproductive parameters and egg contaminant levels of Great Lakes herring gulls. *J. Wildl. Manage.* 41:458-468.

Gilmour, C. C. and E. A. Henry. 1991. Mercury methylation in aquatic systems affected by acid deposition. *Environ. Pollut.* 71:131-169.

Gilmour, C. C., E. A. Henry, and R. Mitchell. 1992. Sulfate stimulation of mercury methylation in freshwater sediments. *Environ. Sci. Technol.* 26:2281-2287.

Gilmour, C. C., G. S. Riedel, M. C. Ederington, J. T. Bell, J. M. Benoit, G. A. Gill, and M. C. Stordal. 1998. Methylmercury concentrations and production rates across a trophic gradient in the northern Everglades. *Biogeochemistry* 40(2-3):327-345.

Goulden, P. D. and D. H. J. Anthony. 1980. Chemical speciation of mercury in natural waters. *Analytica Chimica Acta.* 120:129-139.

Greenfield, B. K., T. R. Hrabik, C. J. Harvey, and S. R. Carpenter. 2001. Predicting mercury levels in yellow perch: use of water chemistry, trophic ecology and spatial traits. *Canadian Journal of Fisheries and Aquatic Sciences* 58:1419-1429.

Grieb, T. M., C. T. Driscoll, S. P. Gloss, C. L. Chofield, G. L. Bowie, and D. B. Porcella. 1990. Factors affecting mercury accumulation in fish in the upper Michigan Peninsula. *Environ. Toxicol. Chem.* 9:919-930.

Guimaraes, J. R. D., M. Meili, O. Malm, and E. M. de Souza Brito. 1998. Hg methylation in sediments and floating meadows of a tropical lake in the Panatal floodplain, Brazil. *Sci Total Environ.* 213:165-175.

Gullvag, B. M., J. P. Aagdal, and B. Eskeland. 1978. A fine structural study of liver (littoral) cells of methylmercury fed Japanese quail (*Coturnix coturnix japonica*). *Acta Pharmacol. Toxicol.* 43(2):93-98.

Hakanson, L. 1980. The quantitative impact of pH, bioproduction and Hg-contamination on the Hg-content of fish (pike). *Environ. Pollut.* (Series B). 1:285-304.

Halbrook, R. S., R. L. Brewer, Jr., and D.A. Buehler. 1999. Ecological risk assessment in a large river reservoir: 7. Environmental contaminant accumulation and effects in great blue heron. *Environ. Toxicol. Chem.* 18:641-648.

Hall, B. D., D. M. Rosenberg, and A. P. Wiens. 1998. Methylmercury in aquatic insects from an experimental reservoir. *Can. J. Aquat. Sci.* 55:2036-2047.

Hanko, E., K. Erne, H. Wanntorp, and K. Borg. 1970. Poisoning in ferrets by tissues of alkyl mercury-fed chickens. *Acta. Vet. Scand.* 11(2):268-282.

Hanten Jr., R.P., R.M. Neumann, S.M. Ward, R.J. Carley, C.R. Perkins, and R. Pirrie. 1998. Relationships between concentrations of mercury in largemouth bass and physical and chemical characteristics of Connecticut lakes. *Trans. Am. Fish. Soc.* 127:807-818.

Hatch, W. R. and W. L. Ott. 1968. Determination of sub-microgram quantities of mercury by atomic absorption spectrophotometry. *Anal. Chem.* 40:2085-2087.

Heinz, G. 1974. Effects of low dietary levels of methylmercury on mallard reproduction. *Bull. Environ. Contam. Toxicol.* 11:386-392.

Heinz, G. 1976. Methylmercury: Second-year feeding effects on mallard reproduction and duckling behavor. *J. Wildl. Manage.* 40:82-90.

Heinz, G. H., and L. N. Locke. 1976. Brain lesions in mallard ducklings from parents fed methylmercury. *Avian Dis.* 20:9-17.

Helander, B., M. Olsson, and L. Reutergardh. 1982. Residue levels of organochlorine and mercury compounds in unhatched eggs and the relationships to breeding success in white-tailed sea eagles *Haliaeetus albicilla* in Sweden. *Holarc. Ecol.* 5:349-366.

Henny, C. J. and G. B. Herron. 1989. DDE, selenium, mercury, and white-faced ibis reproduction at Carson Lake, Nevada (USA). *J. Wildl. Manage.* 53(4):1032-1045.

Henriksson, K., E. Karppanen, and M. Helimin. 1966. High residue of mercury in Finnish white-tailed eagles. *Ornic Fennica.* 43(2):38-45.

Henry, K. S., K. Kannan, B. W. Nagy, N. R. Kevern, M. J. Zabik, and J. P. Giesy. 1998. Concentrations and hazard assessment of organochlorine contaminants and mercury in smallmouth bass from a remote lake in the upper peninsula of Michigan. *Arch. Environ. Contam. Toxicol.* 34:81-86.

Herrin, R. T., R. C. Lathrop, P. R. Gorski, and A. W. Andren. 1998. Hypolimnetic methylmercury and its uptake by plankton during fall destratification: a key entry point of mercury into lake food chains? *Limnol. Oceanogr.* 43:1476-1486.

Hill, C. H. and J. H. Soares, Jr. 1984. Subchronic mercury exposure in Coturnix and a method of hazard evaluation. *Environ. Toxicol. Chem.* 3:489-502.

Hill, C.H. and J.H. Soares, Jr. 1987. Oral and intramuscular toxicity of inorganic and organic mercury chloride to growing quail. *J. Toxicol. Environ. Health.* 20(1-2):105-116.

Hintelmann, H. 1999. Comparison of different extraction techniques used for methylmercury analysis with respect to accidental formation of methylmercury during sample preparation. *Chemosphere* 39:1093-1105.

Honda, K., J. E. Marcovecchio, S. Kan, R. Tatsukawa, and H. Ogi. 1990. Metal concentrations in pelagic seabirds from the North Pacific Ocean. *Arch. Environ. Contam. Toxicol.* 19(5):704-711.

Horvat, M., N. S. Bloom, and L. Liang. 1993. Comparison of distillation with other current isolation methods for the determination of methylmercury compounds in low level environmental samples (Part I. Sediments). *Analytica Chimica Acta* 281:135-152.

Horvat, M., L. Liang, and N. S. Bloom. 1993. Comparison of distillation with other current isolation methods for the determination of methylmercury compounds in low level environmental samples (Part II. Water). *Analytica Chimica Acta* 282:153-168.

Howells, G. 1990. *Acid Rain and Acid Waters.* Toronto, Canada: Ellis Horwood Ltd.

Huckabee, J. W., J. W. Elwood, and S. B. Hildebrand. 1979. Accumulation of mercury in freshwater biota. Chap. 12 in *The Biogeochemistry of Mercury in the Environment*. Amsterdam: Elsevier/North-Holland Biomedical Press.

Hughes, K. D., P. J. Ewins, and K. E. Clark. 1997. A comparison of mercury levels in feathers and eggs of osprey (*Pandion haliaetus*) in the North American Great Lakes. *Arch. Environ. Contam. Toxicol.* 33:441-452.

Hunt, C. D. 1998. Accuracy in analysis: importance of clean metal sampling and analysis. In *Metals in Surface Waters*, edited by H. E. Allen, A.W. Garrison, and G. W. Luther. Chelsea, MI: Ann Arbor Press.

Hurley, J. P., D. P. Krabbenhoft, L. B. Cleckner, M. L. Olson, G. R. Aiken, and P. S. Rawlik, Jr. 1998. System controls on the aqueous distribution of mercury in the northern Florida Everglades. *Biogeochemistry* 40:293-311.

Inza, B., F. Ribeyre, R. Maury-Brachet, and A. Boudou. 1997. Tissue distribution of inorganic mercury, methylmercury and cadmium in the Asiatic clam (*Corbicula fluminea*) in relation to the contamination levels of the water column and sediment. *Chemosphere* 35:2817-2836.

Jefferies, D. S., D. L. Wales, J. R. M. Kelso, and R. A. Linthurst. 1986. Regional chemical characteristics of lakes in North America: Part I - Eastern Canada. *Water Air Soil Pollut.* 31: 551-567.

Jernelov, A., L. Ndner, and T. Rsson. 1975. Swedish perspectives on mercury pollution. *J. Water Pollut. Conf.* 47:810-822.

Jernelov, A., A-H. Johansson, L. Sorensen, and A. Svenson. 1976. Methylmercury degradation in mink. *Toxicology* 6:315-321.

Kerr D. R. and J. P. Meador. 1996. Modeling dose response using generalized linear models. *Environ. Toxicol. Chem.* 15:395-401.

Kettles, I. M., W. W. Shilts, and W. B. Coker. 1991. Surficial geochemistry, south-central Canadian Shield: implications for environmental assessment. *J. Geochem. Explor.* 41:29-57.

Kim, J. P. and S. Burggraaf. 1999. Mercury bioaccumulation in rainbow trout (*Oncorhynchus mykiss*) and the trout food web in Lakes Okareka, Okaro, Tarawera, Rotomahana, and Rotorua, New Zealand. *Water, Air, Soil Pollut.* 115:535-546.

King, K. A., T. W. Custer, and J. S. Quinn. 1991. Effects of mercury, selenium and organochlorine contaminants on reproduction of Forster's terns and black skimmers nesting in a contaminated Texas Bay. *Arch. Environ. Contam. Toxicol.* 20(1):32-40.

Kitamura, S. 1968. *Determination on mercury content in bodies of inhabitants, cats, fishes, and shells in Minamata District and in the mud on Minamata Bay.* Kumamoto Univ., Japan. pp. 257-266.

Koeman, J. H., M. C. ten Noever de Brauw, and R. H. de Vos. 1969. Chlorinated biphenyls in fish, mussels and birds from the River Rhine and the Netherlands coastal area. *Nature.* 221(5186):1126-1128.

Koeman, J. H., J. Garssen-Hoekstra, E. Pels, and J. J. M. de Goeij. 1971. Poisoning of birds of prey by methylmercury compounds. *Meded. Rijksfac. Landbouwwet. Gent.* 36:43-49.

Koeman, J. H., R. H. Hadderingh, and M. F. I. J. Bijlveld. 1972. Persistent pollutants in the white-tailed eagle (Haliaeetus albicilla) in the Federal Public of Germany. *Biol. Conserv.* 4: 373-377.

Koivusaari, J., I. Nuuja, R. Palokangas, and M. Finnlund, M. 1980. Relationships between productivity, eggshell thickness and pollutant contents of addled eggs in the population of white-tailed eagles *Haliaetus albicilla L.* in Finland during 1969-1978. *Environmental Pollution* 23:41-52.

Krabbenhoft, D. P., J. G. Weiner, W. G. Brumbaugh, M. L. Olson, J. F. DeWild, and T. J. Sabin. 1999. A National Pilot Study of Mercury Contamination of Aquatic Ecosystems along Mulitple Gradients. *U.S. Geological Survey Toxic Substances Hydrology Program – Proceedings of the technical meeting, Charleston, S.C., March 8-12, 1999.* USGS. Water Resources Investigations Report 99-4018B, Vol. 2, p147-160.

Kruuk, H. and J. W. H. Conroy. 1991. Mortality of otters (*Lutra lutra*) in Shetland. *J. Appl. Ecol.* 28:83-94.

Lee, Y.H. and A. Iverfeldt. 1991. Measurement of methylmercury and mercury in run-off, lake and rain waters. *Water Air Soil Pollut.* 56:309-321.

Lee, T. H. and S. J. Jiang. 2000. Determination of mercury compounds by capillary electrophoresis inductively coupled plasma mass spectrometry with microconcentric nebulization. *Analytica Chimica Acta* 413:197-205.

Liang, L., M. Horvat, and N. S. Bloom. 1994. An improved speciation method for mercury by GC/CVAFS after aqueous phase ethylation and room temperature precollection. *Talanta.* 41:371-379.

Lindberg, S. E. and W. J. Stratton. 1998. Atmosperhic mercury speciation: concentrations and behavior of reactive gaseous mercury in ambient air. *Environ. Sci. Technol.* 332:49-57.

Lo, J. M. and C. M. Wai. 1975. Mercury loss from water during storage: Mechanisms and prevention. *Anal. Chem.* 47:1869-1870.

Lockhart, W. L., J. F. Uthe, A. R. Kenney, and P. M. Mehrle. 1972. Methylmercury in northern pike (*Esox lucius*), distribution, elimination and biochemical characteristics of contaminated fish. *J. Fish. Res. Board Can.* 29:1519-1523.

Lu, X., W. K. Johnson, and C. S. Wong. 1986. Seasonal replacement of mercury in a coastal fjord by its intermittent anoxicity. *Mar. Poll. Bull.* 17:263-267.

McFarren, E. F., R. J. Lishka, and J. H. Parker. 1970. Criteria for judging acceptability of analytical methods. *Anal. Chem.* 42:358-364.

McKim, M. W., G. F. Olsen, G. W. Holcombe, and E. P. Hunt. 1976. Long-term effects of methylmercuric chloride on three generations of brook trout (*Salvelinus fontnicalis*): toxicity, accumulation, distribution, and elimination. *J. Fish. Res. Board Can.* 33:2726-2931.

McMurtry, M. J., D. L. Wales, W. A. Scheider, G. L. Beggs and P. E. Dimond. 1989. Relationship of mercury concentrations in lake trout (*Salvelinus namaycush*) and smallmouth bass (*Micropterus dolomieui*) to the physical and chemical characteristics of Ontario lakes. *Can. J. Fish. Aquat. Sci.* 46:426-434.

Marsh, R. C. 1980. Federal standard 209 – An effective clean room standard. *J. Environ. Sci.* 23:15-17.

Marsh, R.C. 1981. Federal standard 209 – An effective clean room standard. *J. Environ. Sci.* 24:38-41.

Matida, Y., H. Kumada, S. Kimura, Y. Saiga, T. Nose, M. Yokote, and H. Kawatsu. 1971. Toxicity of mercury compounds to aquatic organisms and accumulation of the compounds by the organisms. *Bull. Freshwater Fish. Res. Lab. Jpn.* 21.

Meyer, M. W. 1998. Ecological risk of mercury in the environment: the inadequacy of "the best available science". *Environ. Toxicol. Chem.* 17:137-138.

Meyer, M. W., D. C. Evers, J. J. Hartigan, and P. S. Rasmussen. 1998. Patterns of common loon (*Gavia immer*) mercury exposure, reproduction, and survival in Wisconsin, USA. Environ. *Toxicol. Chem.* 17:184-190.

Mierle G. and R. Ingram. 1991. The role of humic substances in the mobilization of mercury from watersheds. *Water Air Soil Pollut.* 56:349-357.

Minagawa, K., Y. Takizawa, and I. Kifune. 1980. Determination of very low levels of inorganic and organic mercury in natural waters by cold-vapor atomic absorption spectrometry after preconcentration on a chelating resin. *Analytica Chimica Acta.* 115:103-110.

Miskimmin, B. M., J. W. M. Rudd, and C. A. Kelly. 1992. Influence of dissolved organic carbon, pH, and microbial respiration rates on mercury methylation and demethylation in lake water. *Can. J. Fish. Aquat. Sci.* 49:17-22.

Mitchell J. L. 1961. Mink movements and populations on a Montana river. *J Wildl Manage* 25:48-54.

Molot, L. A., P. J. Dillon, and G. M. Booth. 1990. Whole-lake and nearshore water chemistry in Bowland Lake, before and after treatment with $CaCO_3$. *Can. J. Fish. Aquat. Sci.* 47:412-421.

Monson, B. A. and P. L. Brezonik. 1998. Seasonal patterns of mercury species in water and plankton from softwater lakes in northeastern Minnesota. *Biogeochem.* 40:147-162.

Monson, B.A. and P.L. Brezonik. 1999. Influence of food, aquatic humus and alkalinity on methylmercury uptake by *Daphnia magna. Environ. Toxicol. Chem.* 18:560-566.

Moore D. R. J., R. L. Breton, K. Lloyd. 1997. The effects of hexachlorobenzene on mink in the Canadian environment: An ecological risk assessment. *Environ. Toxicol. Chem.* 16:1042-1050.

Moore, D. R. J., B. Sample, G. Suter, B. Parkhurst, and R. S. Teed. 1999a. A probabilistic risk assessment of the effects of methylmercury and PCBs on mink and kingfishers along East Fork Poplar Creek, Oak Ridge, Tennessee. *Environmental Toxicology and Chemistry* 18(12):2941-2953

Moore, D. R. J., B. Sample, G. Suter, B. Parkhurst, and R. S. Teed. 1999b. Risk-based decision-making: The East Fork Poplar Creek case study. *Environmental Toxicology and Chemistry* 18(12):2954-2958.

Morel, F. M., A. M. Kraepiel, and M. Amyot. 1998. The chemical cycle and bioaccumulation of mercury. *Annu. Rev. Ecol. Syst.* 29:543-566.

Munn, M. D. and T. M. Short. 1997. Spatial heterogeneity of mercury bioaccumulation by walleye in Franklin D. Roosevelt Lake and the upper Columbia River, Washington. Trans. *Amer. Fish. Soc.* 126:477-487.

Mushak, P. 1987. The Quantitative Measurement of Methylmercury. In *Toxicity of Methylmercury*, edited by C.U. Eccles and Z. Annau. Baltimore, Md.: The Johns Hopkins University Press.

Nagase, H., Y. Ose, T. Sato, and T. Ishikawa. 1982. Methylation of mercury by humic substances in an aquatic environment. *Sci. Total Environ.* 24:133-142.

Nagy K. A. 1987. Field metabolic rate and food requirement scaling in mammals and birds. *Ecol. Monogr.* 57:111-128.

NRCC (National Research Council of Canada). 1981. *Acidification in the Canadian aquatic environment: scientific criteria for assessing the effects of acidic deposition on aquatic ecosystems.* Associate Committee on Scientific Criteria for Environmental Quality. NRCC Report No. 18475.

Neary, B. P., P. J. Dillon, J. R. Munro, and B. J. Clark. 1990. *The acidification of Ontario lakes: an assessment of their sensitivity and current status with respect to biological damage.* Toronto, Canada: Ontario Ministry of Environment.

Neumann, R. M. and S. M. Ward. 1999. Bioaccumulation and biomagnification of mercury in two warmwater fish communities. *J. Freshwater Ecol.* 14:487-497.

Newton, I. and M. Haas. 1988. Pollutants in merlin eggs and their effects on breeding. *Br. Birds* 81(6):258-269.

Newton, I., J. A. Bogan, and M. B. Haas. 1989. Organochlorines and mercury in the eggs of British peregrines *Falco peregrinus. Ibis.* 131(3):355-376.

Newton, I. and E. A. Galbraith. 1991. Organochlorines and mercury in the eggs of Golden Eagles, *Aquila chrysaetos*, from Scotland (UK). *Ibis*. 133(2):115-120.

Nichols, J. A. and L. R. Hageman. 1979. Noncontaminating, representative sampling by shattering of cold, brittle, biological tissues. *Anal. Chem.* 51:1591-1593.

Nicholson, J. K. and D. Osborn. 1984. Kidney lesions in juvenile starlings *Sturnus vulgaris* fed on a mercury-contaminated synthetic diet. *Environ. Pollut. Ser.* 33:195-206.

Niimi, A. J. and L. Lowe-Jinde. 1984. Differential blood cell ratios of rainbow trout (*Salmo gairdneri*) exposed to methylmercury and chlorobenzenes. *Arch. Environ. Contam. Toxicol.* 13:303-311.

Niimi, A. J. and G. P. Kissoon. 1994. Evaluation of the critical body burden concept based on inorganic and organic mercury toxicity to rainbow trout (*Oncorhynchus mykiss*). *Arch. Environ. Contam. Toxicol.* 26:169-178.

Norheim, G. and A. Froslic. 1978. The degree of methylation and organ distribution in some birds of prey in Norway. *Acta. Pharmacol. Toxicol.* 43:196-204.

Nuutinen, S. and J. V. K. Kukkonen. 1998. The effect of selenium and organic material in lake sediments on the bioaccumulation of methylmercury by *Lumbricus variegatus* (ologochaeta). *Biogeochem.* 40:267-278.

O'Connor, D. J. and S. W. Nielson. 1980. Environmental survey of methylmercury levels in wild mink (*Mustela vison*) and otter (*Lutra canadensis*) from the northeastern United States and experimental pathology of methylmercurialism in the otter. *Worldwide Furbearer Conference: Proceedings.* Frostburg, MD.

Odin, M., F. Ribeyre, and A. Boudou. 1997. Depuration processes after exposure of burrowing mayfly nymphs (*Hexagenia rigida*) to methylmercury and cadmium from water column or sediment:Effects of temperature and pH. *Aquatic Toxicol.* 37:125-137.

O'Driscoll, N., T. A. Clair, and A. Renca, eds. 2001. *Cycling of Mercury in Kejimkujik National Park: Toxic Substance Research Initiative Project #124 Summary, December 2001 Occasional Report no. 18.* Environment Canada: Sackville, New Brunswick. 93p.

Odsjo, T. and J. Sondell. 1977. Population development and breeding success in the marsh harrier *Circus aeruginosis* in relation to levels of DDT, PCB, and mercury. *Var Fagelvarld.* 36:152-160.

Oehme, G. 1981. Zur Quecksilberruckstandsbelastung tot aufgefundener Seeadler, *Haliaeetus albicilla,* in den Jahren 1967-1978. *Hercynia.* 18:353-364.

Olson, G. F., D. I. Mount, V. M. Snarski, and T. W. Thorslund. 1975. Mercury residues in fathead minnows, *Pimephales promelas* Rafinesque, chronically exposed to methylmercury in water. *Bull. Environ. Contam. Toxicol.* 14:129-134.

Padberg, S., A. Iverfeldt, Y-H. Lee, F. Baldi, M. Filippelli, K. May, and M. Stoeppler. 1994. Determination of low-level methylmercury concentrations in water ans complex matrices by different analytical methods: a methodological intercomparison. In *Mercury Pollution Integration and Synthesis*, edited by C.J. Watras and J.W. Huchabee. Boca Raton, FL: CRC Press.

Palmieri, H. E. L. and L. V. Leonel. 2000. Determination of methylmercury in fish tissue by gas chromatography with microwave-induced plasma atomic emission spectrometry after derivatization with sodium tetraphenylborate. *Fresenius. J. Anal. Chem.* 366:466-469.

Palmisano, F., N. Cardellicchio, and P. G. Zambonin. 1995. Speciation of mercury in dolphin liver: a two-stage mechanism for the demethylation accumulation process and role of selenium. *Mar. Environ. Res.* 40:109-121.

Park, J. G. and L. R. Curtis. 1997. Mercury distribution in sediments and bioaccumulation by fish in two Oregon reservoirs: Point-source and nonpoint-source impacted systems. *Arch. Environ. Contam. Toxicol.* 33:423-429.

Parkhurst, C. R. and P. Thaxton. 1973. Toxicity of mercury to young chickens. *Poult. Sci.* 52:273-276.

Peakall, D. B. and J. L. Lincer. 1972. Methylmercury: Its effect on eggshell thickness. *Bull. Environ. Contam. Toxicol.* 8:89-90.

Rada, R. G., D. E. Powell, and J. G. Wiener. 1993. Whole-lake burdens and spatial distribution of mercury in surficial sediments in Wisconsin seepage lakes. *Can. J. Fish. Aquat. Sci.* 50: 865-873.

Rasmussen, J. B., L. Godbout, and M. Schallenberg. 1989. The humic content of lake water and its relationship to watershed and lake morphometry. *Limnol. Oceanogr.* 34:1336-1343.

Rawson, D. S. 1952. Mean depth and the fish production of large lakes. *Ecology* 33:513-521.

Richardson, G. M. 1994. Physical, Chemical and Geochemical Factors Influencing Mercury Accumulation in Freshwater Fish and Humans in Ontario, Canada. Thesis submitted to the School of Graduate Studies and Research, University of Ottawa, in partial fulfilment of the requirements for the degree of Doctor of Science. 195p.

Richardson, G. M. and D. J. Currie. 1995. Using empirical methods to assess the risks of mercury accumulation in fish from lakes receiving acid rain. *Human and Ecological Risk Assessment* 1:306-322.

Richardson, G. M., M. Egyed, and D. J. Currie. 1995a. Does acid rain increase human exposure to mercury? A review and analysis of recent literature. *Environ. Toxicol. Chem.* 14:809-813.

Richardson, G. M., M. Egyed, and D. J. Currie. 1995b. Human exposure to mercury may decrease as acidic deposition increases. *Water Air Soil Pollut.* 80:31-39.

Richardson, G. M. and D. Burmaster. 1996. Theoretical, toxicological and biostatistical foundations for deriving probability density functions describing variability and uncertainty in references doses, benchmark doses and slope factors. *Human and Ecological Risk Assessment* 2(1):59-61.

Richman, L. A., C. D. Wren, and P. M. Stokes. 1988. Facts and fallacies concerning mer-cury uptake by fish in acid stressed lakes. *Water, Air, Soil, Pollut.* 37:465-473.

Richter, E. M., M. A. Augelli, G. H. Kume, R. N. Mioshi, and L. Angnes. 2000. Gold electrodes from the recordable CDs for mercury quantification by flow injection analysis. *Fresenius J. Anal. Chem.* 366:444-448.

Roberston, D. E., D. S. Sklarew, K. B. Olsen, N. S. Bloom, E. A. Crecelius, and C. W. Apts. 1987. *Measurement of bioavailable mercury species in fresh water and sediments.* Electric Power Research Institute, Inc.

Rodgers, D. W. and F. W. H. Beamish. 1982. Dynamics of dietary methylmercury in rainbow trout, Salmo gairdneri. *Aquat. Toxicol.* 2:271-290.

Rodgers, D. W. and F. W. H. Beamish. 1983. Water quality modifies uptake of waterborne methylmercury by rainbow trout, Salmo gairdneri. *Can. J. Fish. Aquat. Sci.* 40:824-828.

Roelke, M. E., D. P. Schultz, C. F. Facemire, S. F. Sundlof, and H. E. Royals. 1991. *Mercury contamination in Florida panthers.* Gainseville, FL: Florida Panther Interagency Committee.

Rose, J., M. S. Hutcheson, C. R. West, O. Pancorbo, K. Hulme, A. Cooperman, G. DeCe-sare, R. Isaac, and A. Screpetis. 1999. Fish mercury distribution in Massachusetts, USA lakes. *Environ. Toxicol. Chem.* 18:1370-1379.

Ryder, R. A. 1964. Chemical characteristics of Ontario lakes as related to glacial history. *Trans. Am. Fish. Soc.* 93:260-268.

Ryder, R. A. 1965. A method for estimating the potential fish production of north-temperate lakes. *Trans. Am. Fish. Soc.* 94:214-218.

Sanchez-Pedreno, C., J. A. Ortuno, M. I. Albero, M. S. Garcia, and M. V. Valero. 2000. Development of a new bulk optode membrane for the determination of mercury(II). *Analytica Chimica Acta.* 414:195-203.

Sanemasa, I., T. Deguchi, K. Urata, J. Tomooka, and H. Nagai. 1976. Loss and recovery of mercury from sea water during storage. *Analytica Chimica Acta.* 87:479-481.

Sanemasa, I., T. Deguchi, K. Urata, J. Tomooka, and H. Nagai. 1977. The effect of ammonium thiocyanate and sodium chloride on loss and recovery of mercury from water during storage. *Analytica Chimica Acta.* 94:421-427.

SAS Institute. 1990. *SAS/STAT User's Guide*, Version 6, Fourth Edition, Volume 2. Cary, NC: SAS Institute Inc. 846 pp.

Scheider, W. A., D. S. Jefferies, and P. J. Dillon. 1979. Effects of acidic precipitation on precambrian freshwaters in southern Ontario. *J. Great Lakes Res.* 5:45-51.

Scherer, E., F. A. J. Armstrong, and S. H. Nowak. 1975. Effects of mercury contaminated diet upon walleyes, *Stizostedion vitreum vitreum* (Mitchill). *Can. Fish. Mar. Serv. Resour. Dev. Branch Winnipeg Tech. Rep. No. 597.* Winnipeg, Canada. 21pp.

Scheuhammer, A. M., C. M. Atchison, A. H. K. Wong, and D. C. Evers. 1998. Mercury exposure in breeding common loons (*Gavia immer*) in central Ontario, Canada. *Environ. Toxicol. Chem.* 17:191-196.

Scheuhammer, A. M. and J. E. Graham. 1999. The bioaccumulation of mercury in aquatic organisms from two similar lakes with differing pH. *Ecotoxicol.* 8:49-56.

Schindler, D. W. 1971. A hypothesis to explain differences and similarities among lakes in the Experimental Lakes Area, northwestern Ontario. *J. Fish. Res. Bd. Can.* 28:295-301.

Schindler, D. W., S. E. Bayley, P. J. Curtis, B. R. Parker, M .P. Stanton, and C. A. Kelly. 1992. Natural and man-caused factors affecting the abundance and cycling of dissolved organic substances in precambrian shield lakes. *Hydrobiologia*. 229:1-21.

Schroeder, W. H., R. Ebinbhaus, M. Shoeib, K. Timoschenko, and L. A. Barrie. 1995. Atmospheric mercury measurements in the northern hemisphere from 56° to 82.5° N latitude. *Water, Air, Soil Pollut.* 80:1227-1236.

Scott, M. L., J. R. Zimmermann, S. Marinsky, P. A. Mullenhoff, G. L. Rumsey, and R. W. Rice. 1975. Effects of PCBs, DDT, and mercury compounds upon egg production, hatchability and shell quality in chickens and Japanese quail. *Poult Sci.* 54(2):350-368.

Shilts, W. W. and I. M. Kettles. 1989. Geology and acid rain in eastern Canada. *GEOS*. 18:25-32.

Simon, O., F. Ribeyre, and A. Boudou. 2000. Comparative experimental study of cadmium and methylmercury trophic transfers between the Asiatic clam *Corbicula fluminea* and the crayfish *Astacus astacus. Arch. Environ. Contam. Toxicol.* 38:317-326.

Skoog, D. A., F. J. Holler, and T. A. Nieman, eds. 1998. *Principles of instrumental analysis: fifth edition.* Orlando, FL: Harcourt Brace & Company.

Slemr, F., G. Schuster, and W. Seiler. 1985. Distribution, speciation, and budget of atmospheric mercury. *J. Atmos. Chem.* 3:407-434.

Snodgrass, J. W., C. H. Jagoe, A. L. Bryan, Jr., H. A. Brant, and J. Burger. 2000. Effects of trophic status and wetland morphology, hydroperiod, and water chemistry on mercury concentrations in fish. *Can. J. Aquat. Sci.* 57:171-180.

Sorensen, J. A., G. E. Glass, K. W. Schmidt, J. K. Huber, and G. R. Rapp Jr. 1990. Airborne mercury deposition and watershed characteristics in relation to mercury concentrations in water, sediments, plankton, and fish of eighty northern Minnesota lakes. *Environ. Sci. Technol.* 24(11):1716-1727.

Southworth, G. R., M. J. Peterson, and M.G. Ryon. 2000. Long-term increased bioaccumulation of mercury in largemouth bass follows reduction of waterborne selenium. *Chemosphere*. 41:1101-1105.

Southworth, G. R., R. R. Turner, M. J. Peterson, and M. A. Bogle. 1995. Form of mercury in stream fish exposed to high concentrations of dissolved inorganic mercury. *Chemosphere* 30:779-787.

Spann, J. W., R. G. Heath, J. F. Kreitzer, and L. N. Locke. 1972. Ethyl mercury p-toluene sulfonanilide: lethal and reproductive effects on pheasants. *Science* 175:328-331.

Spry, D. J. and J. G. Wiener. 1991. Metal bioavailability and toxicity to fish in low-alkalinity lakes: A critical review. *Environ. Pollut.* 71:243-304.

Spuler, S., M. Linne, A. Sappey, and S. Snyder. 2000. Development of a cavity ringworm laser absorption spectrometer for detection of trace levels of mercury. *Applied Optics.* 39:2480-2486.

Stein, E. D., Y. Cohen, andA. M. Wiener. 1996. Environmental distribution and transformation of mercury compounds. *Crit. Rev. Environ. Sci. Technol.* 26:143.

Stoeppler, M. and W. Matthes. 1978. Storage behaviour of inorganic mercury and methylmercury chloride in sea water. *Analytica Chimica Acta* 98:389-392.

Stoewsand, G. S., J. L. Anderson, W. H. Gutenmann, C. A. Bache, and D. L. Lisk. 1971. Eggshell thinning in Japanese quail fed mercuric chloride. *Science* 173:1030-1031.

Sullivan, T. J., J. M. Eilers, M. R. Church, D. J. Blick, K. N. Eshleman, D. H. Landers, and M. S. De Haan. 1988. Atmospheric wet sulphate deposition and lakewater chemistry. *Nature* 331:607-609.

Suter, G. W. 1996. Abuse of hypothesis testing statistics in ecological risk assessment. Hum. *Eco.l Risk Assess.* 2:331-347.

Svobodová, Z., L. Dušek, M. Hejtmánek, B. Vykusová, and R. Šmíd. 1999. Bioaccumulation of mercury in various fish species from Orlík and Kamšk water reservoirs in the Czech Republic. *Ecotoxicol. Environ. Safety* 43:231-240.

Takeuchi, T., H. Worikawa, and Matsumoto. 1962. A pathological study of Minimata disease in Japan. *Acta. Neuropathol.* 2:40-57.

Takeuchi, T., F. M. D'Itri, P. V. Fischer, C. S. Annett, and M. Okabe. 1977. The outbreak of Minimata disease (methylmercury poisoning) in cats on northwestern Ontario *Reserves. Environ. Res.* 13:215-228.

Thaxton, J. P. and C. R. Parkhurst. 1973. Abnormal mating behavior and reproductive dysfunction caused by mercury in Japanese quail. *Proc. Soc. Exper. Biol. Med.* 144:252-255.

Thompson, D. R., K. C. Hamer, and R. W. Furness. 1991. Mercury accumulation in great skuas, Catharacta skua, of known age and sex and its effects upon breeding and survival. *J. Appl. Ecol.* 28(2):672-684.

Thurman, E. M. 1985. *Organic Geochemistry of Natural Waters.* Dordrecht: Martinus Ni-jhoff/Dr. W. Junk Publishers.

Tsubaki, T. and H. Takahashi. 1986. Recent advances in Minamata disease studies. Tokyo, Japan: Kodanska, Ltd.

U.S. EPA (United States Environmental Protection Agency). 1985. *Ambient Water Quality Criteria for Mercury - 1984.* PB85-227452. Washington, DC: Office of Water, United States Environmental Protection Agency.

U.S. EPA (United States Environmental Protection Agency). 1993a. *Great Lakes Water Quality Initiative Criteria Documents for the Protection of Wildlife (Proposed) DDT; Mercury; 2,3,7,8_TCDD; PCBs.* EPA/882/R-93/006. Washington, DC: Office of Science and Technology.

U.S. EPA (United States Environmental Protection Agency). 1993b. Water Quality Guidance for the Great Lakes System and Correction: Proposed Rules. *Fed. Reg.* 58:20802-21047.

U.S. EPA (United States Environmental Protection Agency). 1993c. *Wildlife Exposure Factors Handbook Volume 1.* EPA/600/R-93/187a. Washington, DC: Office of Research and Development, United States Environmental Protection Agency.

U.S. EPA (United States Environmental Protection Agency). 1995. *Great Lakes Water Quality Initiative Criteria Documents for the Protection of Wildlife: DDT; Mercury; 2,3,7,8-TCDD; PCBs., March 1995.* EPA 820-8-96-001. Washington, DC: Office of Water.

U.S. EPA (United States Environmental Protection Agency). 1996a. *Method 1631 Mercury in water by oxidation, purge and trap, and CVAFS.* Washington, DC: U.S. Environmental Protection Agency.

U.S. EPA (United States Environmental Protection Agency). 1996b. *Water Quality Criteria Documents for the Protection of Aquatic Life in Ambient Water.* EPA 820-8-96-001. Washington, DC: Office of Water.

U.S. EPA (United States Environmental Protection Agency). 1997a. *Mercury Study Report to Congress. Volume VI: An Ecological Assessment for Anthropogenic Mercury Emissions in the United States.* EPA-452/R-97-008.

U.S. EPA (United States Environmental Protection Agency). 1997b. *Mercury Study Report to Congress. Volume VII: Characterization of Human Health and Wildlife Risks from Mercury Exposure in the United States.* EPA-452/R-97-009.

U.S. EPA (United States Environmental Protection Agency). 2001. *Method 1631 Revision C: Mercury in Water by Oxidation, Purge and Trap, and Cold Vapour Atomic Fluorescence Spectrometry.* EPA-821-R-01-024.

Uthe, J. F., J. Solomon, and B. Grift. 1972. Rapid semimicro method for the determination of methylmercury in fish tissue. *Association of Official Analytical Chemists Journal* 55(3):583-589.

Vermeer, K., F. A. J. Armstrong, and D. R. M. Hatch. 1973. Mercury in aquatic birds at Clay Lake, western Ontario. *J. Wildl. Manage.* 37(1):58-61.

Wang, J., Z. Xiao, and O. Lindqvist. 1995. On-line measurement of mercury in simulated flue gas. *Water, Air, Soil Pollut.* 80:1217-1226.

Watras, C. J., R. C. Back, S. Halvorsen, R. J. M. Hudson, K. A. Morrison, and S. P.Wente. 1998. Bioaccumulation of mercury in pelagic freshwater food webs. *Sci. Total Environ.* 219:183-208.

Weber, J.H. 1993. Review of possible paths for abiotic methylation of mercury (II) in the aquatic environment. *Chemosphere* 26:2063-2077.

Weiss, H. V., W. H. Shipman, and M. A. Guttman. 1976. Effective storage of dilute mercury solutions in polyethylene. *Analytica Chimica Acta.* 81:211-217.

Welsh, S. O., and J. J. Soares. 1976. The protective effect of vitamin E and selenium against methylmercury toxicity in the Japanese quail. *Nutr. Rep. Int.* 13:43-51.

Westoo, G. 1966. Determination of methylmercury compounds in foodstuffs I: methylmercury compounds in fish, identification and determination. *Acta Chemica Scandinavica* 20:2131-2137.

Westoo, G. 1967. Determination of methylmercury compounds in foodstuffs II: determination of methlymercury in fish, egg, meat and liver. *Acta Chem. Scand.* 21(7):1790-1800.

Whitney, S. D. 1991. Effects of maternally transmitted mercury on the hatching success, survival, growth, and behaviour of embryo and larval walleye (*Stizostedion vitreum vitreum*). Master's Thesis, University of Wisconsin-La Crosse, La Crosse.

Wiemeyer, S. N., T. G. Lamont, C. M. Bunck, C. R. Sindelar, F. J. Gramlich, J. D. Fraser, and M. A. Byrd. 1984. Organochlorine pesticide, polychlorobiphenyl and mercury residues in bald eagle eggs, 1969-79, and their relationships to shell thinning and reproduction. *Arch. Environ. Contam. Toxicol.* 13(5):529-550.

Wiener, J. G., R. E. Martini, T. B. Shefey, and G. E. Glass. 1990. Factors influencing mercury concentrations in walleyes in Northern Wisconsin lakes. *Trans. Am. Fish. Soc.* 119:862-870.

Wiener, J. G. and D. J. Spry. 1996. Toxicological significance of mercury in freshwater fish. Chap. 9 in *Environmental contaminants in wildlife: interpreting tissue concentrations,* edited by W. N. Beyer, G. H. Heinz and A. W. Redmon-Norwood. New York: CRC Lewis Publishers.

Wilken, R.D. and R. Falter. 1998. Determination of methylmercury by the species-specific isotope addition method using a newly developed HPLC-ICP MS coupling technique with ultrasonic nebulization. *Applied. Organometallic. Chem.* 12:551-557.

Winfrey, J. R. and J. W. M. Rudd. 1990. Environmental factors affecting the formation of methylmercury in low pH lakes. *Environ. Toxicol. Chem.* 9:853-869.

Wobeser, G. 1973. Aquatic mercury pollution: studies of its occurrence and pathologic effects on fish and mink. Ph.D. thesis, University of Saskatchewan.

Wobeser, G. 1975. Acute toxicity of methylmercury chloride and mercuric chloride for rainbow trout (*Salmo gairdneri*) fry and fingerling. *J. Fish. Res. Board Can.* 32:2005.

Wobeser, G. 1976. Mercury poisoning in a wild mink. *Journal of Wildlife Diseases* 12:335-340.

Wobeser, G., N. O. Nielsen, and B. Schiefer. 1976a. Mercury and mink I. Use of mercury-contaminated fish as a food for ranch mink intoxication. *Can. J. Comp. Med.* 40(1):30-33.

Wobeser, G., N. O. Nielsen, and B. Schiefer. 1976b. Mercury and mink II. Experimental methylmercury intoxication. *Can. J. Comp. Med.* 40(1):34-45.

Wolfe, M. and D. Norman. 1998. Effects of waterborne mercury on terrestrial wildlife at Clear Lake: Evaluation and testing of a predictive model. *Environ. Toxicol. Chem.* 17:214-227.

Wren, C. D. 1985. Probable case of mercury poisoning in a wild otter, Lutra canadensis, in northwestern Ontario. *Can. Field. Nat.* 99:112-114.

Wren, C., H. MacCrimmon, R. Frank, and P. Suda. 1980. Total and methylmercury levels in wild mammals from the precambrian shield area south central Ontario, Canada. *Bull. Environ. Contam. Toxicol.* 25(1):100-105.

Wren, C. D., D. B. Hunter, J. F. Leatherland, and P. M. Stokes. 1987a. The effects of polychlorinated biphenyls and methylmercury, singly and in combination, on mink. I: Uptake and toxic responses. *Arch. Environ. Contam. Toxicol.* 16:441-447.

Wren, C. D., D. B. Hunter, J. F. Leatherland, and P. M. Stokes. 1987b. The effects of polychlorinated biphenyls and methylmercury, singly and in combination on mink. 2. Reproduction and kit development. *Arch. Environ. Contam. Toxicol.* 16:449-454.

Wren, C. D., W. A. Scheider, D. L. Wales, B. W. Muncaster, and I. M. Gray. 1991. Relation between mercury concentrations in walleye (*Stizostedion vitreum vitreum*) and northern pike (*Esox lucius*) in Ontario lakes and influence of environmental factors. *Can. J. Fish. Aquat. Sci.* 48:132-139.

Xinbin, F., J. Sommar, K. Gårdfelt, and O. Lindqvist. 2000. Improved determination of gaseous divalent mercury in ambient air using KCl coated denuders. *Fresenius J. Anal. Chem.* 366:423-428.

Xun, L., N. E. R. Campbell, and J. W. M. Rudd. 1987. Measurements of specific rates of net methylmercury production in the water column and surface sediments of acidification and circumneutral lakes. *Can. J. Fish. Aquat. Sci.* 44:750-757.

WASTEWATER UTILITY

Alabama
Montgomery Water Works & Sanitary Sewer Board

Alaska
Anchorage Water & Wastewater Utility

Arizona
Gila Resources
Glendale, City of, Utilities Department
Mesa, City of
Phoenix Water Services Department
Pima County Wastewater Management

Arkansas
Little Rock Wastewater Utility

California
Calaveras County Water District
Central Contra Costa Sanitary District
Contra Costa Water District
Crestline Sanitation District
Delta Diablo Sanitation District
Dublin San Ramon Services District
East Bay Municipal Utility District
El Dorado Irrigation District
Escondido, City of
Fairfield-Suisun Sewer District
Irvine Ranch Water District
Las Virgenes Municipal Water District
Lodi, City of
Los Angeles, City of
Los Angeles County, Sanitation Districts of
Napa Sanitation District
Orange County Sanitation District
Palo Alto, City of
Riverside, City of
Sacramento Regional County Sanitation District
San Diego Metropolitan Wastewater Department, City of
San Francisco, City & County of
San Jose, City of
Santa Barbara, City of
Santa Rosa, City of
South Bayside System Authority
South Orange County Wastewater Authority
Union Sanitary District
West Valley Sanitation District

Colorado
Boulder, City of
Colorado Springs, City of
Littleton/Englewood Water Pollution Control Plant
Metro Wastewater Reclamation District, Denver

Connecticut
New Haven, City of, WPCA

District of Columbia
District of Columbia Water & Sewer Authority

Florida
Broward, County of
Fort Lauderdale, City of
Gainesville Regional Utilities
JEA
Kissimmee, City of, Department of Water Resources
Miami-Dade Water & Sewer Authority
Orange County Utilities Department
Orlando, City of
Reedy Creek Improvement District
Seminole County Environmental Services
St. Petersburg, City of
Stuart Public Utilities
Tallahassee, City of
Tampa, City of
West Palm Beach, City of

Georgia
Atlanta Department of Watershed Management
Augusta, City of
Clayton County Water Authority
Cobb County Water System
Columbus Water Works
Fulton County
Gwinnett County Department of Public Utilities
Macon Water Authority
Savannah, City of

Hawaii
Honolulu, City and County of

Idaho
Pocatello, City of

Illinois
American Bottoms Wastewater Treatment Plant
Dupage, County of
Greater Peoria Sanitary District
Kankakee River Metropolitan Agency
Metropolitan Water Reclamation District of Greater Chicago
North Shore Sanitary District
Wheaton Sanitary District

Indiana
Fort Wayne, City of

Iowa
Ames, City of
Des Moines Metro Wastewater Reclamation Authority

Kansas
Johnson County Unified Wastewater Districts
Unified Government of Wyandotte County/Kansas City, City of

Kentucky
Louisville & Jefferson County Metropolitan Sewer District

Louisiana
Sewerage & Water Board of New Orleans

Maine
Bangor, City of
Portland Water District

Maryland
Anne Arundel County Bureau of Utility Operations
Howard County Department of Public Works
Washington Suburban Sanitary Commission

Massachusetts
Boston Water & Sewer Commission
Upper Blackstone Water Pollution Abatement District

Michigan
Ann Arbor, City of
Detroit, City of
Grand Rapids, City of
Holland Board of Public Works
Lansing, City of
Owosso Mid-Shiawassee County WWTP
Saginaw, City of
Wayne County Department of Environment
Wyoming, City of

Minnesota
Rochester, City of

Missouri
Independence, City of
Kansas City Missouri Water Services Department
Little Blue Valley Sewer District
Metropolitan St. Louis Sewer District

Nebraska
Lincoln Wastewater System

Nevada
Henderson, City of

New Jersey
Passaic Valley Sewerage Commissioners

New York
New York City Department of Environmental Protection

North Carolina
Cary, Town of
Charlotte/Mecklenburg Utilities
Durham, City of
Metropolitan Sewerage District of Buncombe County
Orange Water & Sewer Authority

Ohio
Akron, City of
Butler County Department of Environmental Services
Columbus, City of
Metropolitan Sewer District of Greater Cincinnati
Northeast Ohio Regional Sewer District
Summit, County of

Oklahoma
Tulsa, City of

Oregon
Clean Water Services
Eugene/Springfield Water Pollution Control
Water Environment Services

Pennsylvania
Philadelphia, City of
University Area Joint Authority, State College

South Carolina
Charleston Commissioners of Public Works
Mount Pleasant Waterworks & Sewer Commission
Spartanburg Sanitary Sewer District

WERF SUBSCRIBERS

Tennessee
Cleveland, City of
Knoxville Utilities Board
Murfreesboro Water & Sewer Department
Nashville Metro Water Services

Texas
Amarillo, City of
Austin, City of
Dallas Water Utilities
Denton, City of
El Paso Water Utilities
Fort Worth, City of
Gulf Coast Waste Disposal Authority
Houston, City of
San Antonio Water System
Trinity River Authority

Utah
Salt Lake City Corporation

Virginia
Alexandria Sanitation Authority
Arlington, County of
Fairfax County Virginia
Hampton Roads Sanitation District
Henrico, County of
Hopewell Regional Wastewater Treatment Facility
Loudoun County Sanitation Authority
Lynchburg Regional WWTP
Prince William County Service Authority
Richmond, City of

Washington
Edmonds, City of
Everett, City of
King County Department of Natural Resources
Seattle Public Utilities
Yakima, City of

Wisconsin
Green Bay Metro Sewerage District
Madison Metropolitan Sewerage District
Milwaukee Metropolitan Sewerage District
Racine, City of
Sheboygan Regional Wastewater Treatment
Wausau Water Works

Australia
South Australian Water Corp.
Sydney Water Corp.
Water Corp. of Western Australia

Canada
Greater Vancouver Regional District
Toronto, City of, Ontario

Mexico
Servicios de Agua y Drenaje de Monterrey, I.P.D.

United Kingdom
Yorkshire Water Services Limited

STORMWATER UTLITY

California
Los Angeles, City of, Department of Public Works
Monterey, City of
Santa Rosa, City of

Colorado
Boulder, City of

Georgia
Griffin, City of

Kansas
Overland Park, City of

Kentucky
Louisville & Jefferson County Metropolitan Sewer District

Maine
Portland Water District

North Carolina
Cary, Town of
Charlotte, City of, Stormwater Services

Pennsylvania
Philadelphia, City of

STATE

Fresno Metropolitan Flood Control District, Calif.
Urban Drainage & Flood Control District, Colo.

CORPORATE

ADS Environmental Services
The ADVENT Group Inc.
Alan Plummer & Associates
Alden Research Laboratory
Alpine Technology Inc.
Aquateam–Norwegian Water Technology Centre A/S
BaySaver Inc.
BioVir Inc.
Black & Veatch
Boyle Engineering Corporation
BPR CSO
Brown & Caldwell
Burns & McDonnell
CABE Associates Inc.
The Cadmus Group
Camp Dresser & McKee Inc.
Carollo Engineers Inc.
Carpenter Environmental Associates Inc.
CDS Technologies Inc.
Chemtrac Systems Inc.
CH2M HILL
Clancy Environmental Consultants Inc.
Damon S. Williams Associates, LLC
David L. Sheridan, P.C.
Earth Tech Inc.
Ecolab Water Care Services
EMA Inc.
The Eshelman Company Inc.
Finkbeiner, Pettis, & Strout Inc.
Frontier Geosciences Inc.
ftn Associates Inc.
Gannett Fleming Inc.
GE Betz
Golder Associates Inc.
Greeley and Hansen LLC
The HACH Company
Hazen & Sawyer, P.C.
HDR Engineering Inc.
HNTB Corporation
HydroQual Inc.
Insituform Technologies Inc.
Institute for Environmental Technology & Industry, Korea
Jacobson Helgoth Consultants Inc.
Jason Consultants Inc.
Jordan, Jones, & Goulding Inc.
KCI Technologies Inc.
Kelly & Weaver, P.C.
Kennedy/Jenks Consultants
Komline Sanderson Engineering Corporation
Lawler, Matusky & Skelly Engineers, LLP
Limno-Tech Inc.
Lombardo Associates Inc.
Malcolm Pirnie Inc.
McKim & Creed
MEC Analytical Systems Inc.
MWH
New England Organics
Odor & Corrosion Technology Consultants Inc. (OCTC)
ONDEO Degremont Inc.
Oswald Green, LLC
PA Government Services Inc.
Parametrix Inc.
Parsons
Post, Buckley, Schuh & Jernigan
R&D Engineering/Conestoga Rover & Associates
The RETEC Group
R.M. Towill Corporation
Ross & Associates Ltd.
Royce Technologies
SAIC Maritime Technical Group
Sear-Brown
Sierra Environmental Services
SYNAGRO
Tetra Tech Inc.
Trojan Technologies Inc.
URS Corporation
USFilter NATC
Vortechnics Inc.
Wade-Trim Inc.
Weston Solutions Inc.
Woodard & Curran
WRc/D&B, LLC
WWETCO, LLC

INDUSTRY

American Electric Power
ChevronTexaco Energy Research & Technology Company
The Coca-Cola Company
Dow Chemical Company
DuPont Company
Eastman Kodak Company
Merck & Company Inc.
ONDEO Services
Procter & Gamble Company
PSEG Services Corp.
Severn Trent Services Inc.
Shell Global Solutions
Thames Water Plc
United Water Services LLC

Note: List as of 5/12/03

Board of Directors

Chair
Stephen T. Hayashi
Union Sanitary District

Vice-Chair
James F. Stahl
County Sanitation Districts of Los Angeles County

Secretary
William J. Bertera
Water Environment Federation

Treasurer
Karl W. Mueldener
Kansas Department of Health & Environment

Mary E. Buzby, Ph.D.
Merck & Company Inc.

Gordon R. Garner
Prospect, KY

Lawrence P. Jaworski
Greeley and Hansen

Richard D. Kuchenrither, Ph.D.
Black & Veatch

Alfonso R. Lopez
New York City Department of Environmental Protection

Vernon D. Lucy
ONDEO Degremont Inc.

Richard G. Luthy, Ph.D.
Stanford University

Thomas R. Morgan
Montgomery Water Works & Sanitary Sewer Board

John T. Novak, Ph.D.
Virginia Polytechnic Institute & State University

Lynn H. Orphan
Kennedy/Jenks Consultants

Murli Tolaney
MWH

Executive Director
Glenn Reinhardt

Research Council

Chair
John Thomas Novak, Ph.D
Virginia Polytechnic Institute & State University

Vice-Chair
James Crook, Ph.D.
Water Reuse Consultant

Robert G. Arnold, Ph.D.
University of Arizona, Tucson

Robin L. Autenrieth, Ph.D.
Texas A&M University

Glen T. Daigger, Ph.D.
CH2M HILL

James R. Dartez
Royce Technologies

Geoffrey H. Grubbs
U.S. EPA

Michael D. Jawson, Ph.D.
U.S. Department of Agriculture

Norman E. LeBlanc
Hampton Roads Sanitation District

Drew C. McAvoy, Ph.D.
Procter & Gamble Company

Charles C. Murray
Washington Suburban Sanitary Commission

Margaret H. Nellor, P.E.
County Sanitation Districts of Los Angeles County

Spyros Pavlostathis, Ph.D.
Georgia Institute of Technology

Peter J. Ruffier
Eugene/Springfield Water Pollution Control

Michael W. Sweeney, Ph.D.
Louisville & Jefferson Co. Metro Sewer District

George Tchobanoglous, Ph.D.
Tchobanoglous Consulting

Gary Toranzos, Ph.D.
National Science Foundation

Stormwater Technical Advisory Committee

Chair
Christine Andersen, P.E.
City of Boulder

Vice-Chair
Ben Urbonas, P.E.
Urban Drainage and Flood Control District

Gail B. Boyd
URS Corporation

Larry Coffman
Prince George's County

Doug Harrison
Fresno Metro Flood Control District

Robert E. Pitt, Ph.D., P.E., D.E.E.
University of Alabama

A. Charles Rowney, Ph.D.
Camp Dresser & McKee Inc.

James Wheeler, P.E.
U.S. EPA

WERF Product Order Form

As a benefit of joining the Water Environment Research Foundation, subscribers are entitled to receive one complimentary copy of all final reports and other products. Additional copies are available at cost (usually $10). To order your complimentary copy of a report, please write "free" in the unit price column. WERF keeps track of all orders. If the charge differs from what is shown here, we will call to confirm the total before processing.

Name ____________________ Title ____________________

Organization ____________________

Address ____________________

City __________ State __________ Zip Code __________ Country __________

Phone __________ Fax __________ Email __________

Stock #	Product	Quantity	Unit Price	Total
			Postage & Handling	
			VA Residents Add 4.5% Sales Tax	
			Canadian Residents Add 7% GST	
			TOTAL	

Method of Payment: (All orders must be prepaid.)

❑ Check or Money Order Enclosed

❑ Visa ❑ Mastercard ❑ American Express

Account No. ____________________ Exp. Date __________

Signature ____________________

Shipping & Handling:

Amount of Order	United States	Canada & Mexico	All Others
Up to but not more than:	Add:	Add:	Add:
$20.00	$5.00*	$8.00	50% of amount
30.00	5.50	8.00	40% of amount
40.00	6.00	8.00	
50.00	6.50	14.00	
60.00	7.00	14.00	
80.00	8.00	14.00	
100.00	10.00	21.00	
150.00	12.50	28.00	
200.00	15.00	35.00	
More than $200.00	Add 20% of order	Add 20% of order	

*minimum amount for all orders

Note: Please make checks payable to the Water Environment Research Foundation.

To Order (Subscribers Only):

Log on to www.werf.org and click on the "Product Catalog."

Phone: (703) 684-2470
Fax: (703) 299-0742.

WERF
Attn: Subscriber Services
601 Wythe Street
Alexandria, VA 22314-1994

To Order (Non-Subscribers):

Non-subscribers may be able to order WERF publications either through WEF (www.wef.org) or IWAP (www.iwapublishing.com). Visit WERF's website at www.werf.org for details.

IWA Publishing's authorised EU representative for General Product Safety Regulations is Diane D'Arras, 15 rue Duret, 75116 Paris, France, e-mail: safety@iwap.co.uk.

Printed and bound by CPI Group (UK) Ltd, Croydon, CR0 4YY
12/05/2026
02108912-0002